The Complete Greenhouse Gardener

by the same Author

The Complete Gardener · The Complete Vegetable Grower
The Compost Flower Grower · The Compost Fruit Grower
The ABC of Gardening · The Basic Book of Vegetable Gardening
The ABC of Fruit Growing · The Basic Book of Flower Growing
The ABC of the Greenhouse · The Basic Book of the Rock Garden and Pool
The ABC of Bulbs and Corms · The ABC of Garden Pests and Diseases
The ABC of Continuous Cloches · The ABC of Flowering Shrubs
The ABC of Pruning · The Organic Guide to Soils, Humus and Health
The ABC of Herbaceous Borders · Herbs, Salads and Tomatoes
The Basic Book on Roses · The Basic Book on Chrysanthemums
The ABC of Carnations and Pinks · Cacti as House Plants
The Basic Book on Dahlias · The ABC Guide to Garden Flowers
The Beginner's ABC of Gardening · Town and City Gardening
Mini-Work Gardening · The Horticultural Note Book
The Reason Why of Gardening · Semi-detached Gardening
Etc., Etc.

THE COMPLETE GREENHOUSE GARDENER

W. E. Shewell-Cooper
M.B.E., N.D.H., F.L.S., F.R.S.L., D.LITT., Dip. Hort. (Wye)
Commandeur du Merite Agricole (France)
Knight of Merit (Italy)
Fellow and Doctor of the Horticultural College (Vienna)
Director, The International Horticultural Advisory Bureau
Hon. Director, The Good Gardeners' Association
Lately Command Horticultural Advisor
Eastern and South-Eastern Commands
1940-1949

COLLINS
ST JAMES'S PLACE, LONDON
1976

William Collins Sons & Co Ltd
London · Glasgow · Sydney · Auckland
Toronto · Johannesburg

Dedicated to
my A.T.S. Staff (now W.R.A.C.)
in the Horticultural Department of Eastern Command
Senior Commander I. R. Shewell-Cooper, A.T.S.
Warrant-Officer B.B.M. Brown-Constable, A.T.S.
Sergeant M.E. Lean, A.T.S.
as well as
Miss Gweneth Wood
Miss Marjorie Saunders
Miss Nancy Oliver
and
Mrs. Netta Waller
who acted as Instructors at
the Army Horticultural Training Centres
at Cobham, Maidstone and Colchester
1940-1948

First published 1976

ISBN 0 00 219591 7
Made and Printed in Great Britain by
William Collins Sons & Co Ltd Glasgow

Contents

Colour Photographs

All colour photographs are by Pat Brindley except *Cyperus alternifolius,* which is by Brian Furner.

Black-and-White Photographs

All black-and-white photographs are by Pat Brindley except where otherwise indicated.

Preface

EVER since *The Complete Gardener* was published 22 years ago, I have wanted to write this companion volume on the growing of plants under glass.

Long before becoming Superintendent of the Swanley Horticultural College, before the 1939-45 war, I had had the opportunity of studying greenhouse problems at first hand in many parts of England.

Although the enthusiastic gardener may start with vegetables for the home, then progress to a lawn and beautiful flowers, he will eventually feel the need to put up a greenhouse which may become the central working feature of the garden as he gets older. Many thousands of greenhouses are bought each year all over the country, though by far the greatest number are in the north of Britain.

My grateful appreciation for their advice, assistance and encouragement must go to Mr A. W. Farnfield (ferns), Mr Thomas Rochford (pot-plants), the late Montagu Allwood (carnations), Major Douglas Foxwell, MC (sweet peas), Mr B. G. White (orchids), Mr Christopher Rochford (cacti), Mr Jack Woodman (chrysanthemums); also to Mr C. P. Quarell B.Sc., N.D.H., Mrs Gweneth Johnson, Mr C. G. R. Shewell-Cooper (my son) and Mrs Irene Blum (my secretary) for their hard work.

I hope that this book will be of value to those who intend to grow plants under glass. I shall be glad to hear from readers of any errors which may have crept in.

W. E. Shewell-Cooper

The International Horticultural Advisory Bureau,
Arkley, S. Herts.

PART ONE

The Greenhouse and its Management

CHAPTER I

The Greenhouse

GENERALLY speaking, the ideal material for the greenhouse frame is timber, but some substitutes have been found to possess certain advantages. Steel and aluminium are more durable and their maintenance costs are lower. Because of the strength of metal the supporting members can be made narrower and there is less obstruction to the light.

The modern trend in greenhouse construction is to standardise both size and type of house, as well as all the fittings such as glass, doors, lights, timbers, metal purlins and even bracing struts. Much more, however, should be done in the future to improve the design of the standard greenhouse. The ideal is, of course, to have no internal supports to obstruct light and cultivation, and the greenhouse of the future will almost certainly be without the traditional ridge and heavy ventilators, which at present prevent the entry of a considerable amount of light.

ASPECT. When considering the erection of greenhouses it is necessary to take into account all the natural features of the land chosen. The main consideration is to choose a situation which will give maximum sunlight and yet is sheltered from north and north-east winds. Slight variations in design may be necessary to conform with the land at the disposal of the builder, especially where it is necessary to construct a house in an odd corner of land or where there is an undue slope. In the latter case, for instance, it may be necessary to 'terrace' the house.

The house or houses should normally run north and south to ensure the maximum average sunlight and reduce the shading effect of rafters to the minimum during the course of the year. It is estimated that in winter the light loss due to rafters, doors and dirty glass, can be as high as 50%. Experiments have proved, however, that it is especially important to have houses which are used regularly in the winter running east and west, as this ensures maximum light when the sun is low in the sky.

The optimum angle for house ridges enabling the maximum amount of sunlight to pass through without undue reflection has been found to be 46°, but practical considerations necessitate an angle of about 30–35° being used. In this respect semi-circular polythene houses have an advantage.

It has been suggested that a three-quarter span house facing east and west will allow a greater amount of light to pass through. It is also argued that an increase in glass size from 18 × 20 in. to 24 × 24 in. is an excellent plan, as this reduces the number of rafters necessary.

Materials

TIMBER. This has a great many factors in its favour. It is easily worked and can be cut into the correct shapes and sizes by the use of saws and chisels. It has a high strength in relation to its weight, and does not expand to any great degree during hot weather. There are a few disadvantages, however. Timber will not retain its condition indefinitely, and being a natural product there can be no complete standardisation of pieces. Moreover, it has to be imported because there are very few home-grown woods that are suitable. The types of timber most commonly used are Baltic Red Wood, Red Deal, Douglas Fir and Western Red Cedar. Glazing bars have to be relatively thick and so exclude daylight.

Oak is probably the best material for gutter plating owing to its durability, the only difficulty being that long lengths with straight grain are not often found. It is for this reason that deal is so often used. Certain preservatives will lengthen the life of greenhouse timbers to a considerable degree, e.g. mercuric chloride, copper sulphate, zinc chloride and sodium fluoride, and the wood is dipped for a few minutes in a solution made from any one of them, and then allowed to dry.

Paint is one of the best preservatives. It prevents surface disintegration and, if white, avoids any loss of light because it gives maximum reflection. It also protects the wood from destructive fungi. It need not be used on cedar.

ALUMINIUM ALLOYS. These are the ideal materials, combining great strength and lightness with a lower coefficient of expansion than that of steel. However, the cost is on the whole somewhat higher than that of wood.

STEEL. It is only during the past few years that the use of steel in the manufacture of greenhouse rafters and components has been considered seriously, and the principle followed is similar to that of the steel window frames used in dwelling-houses. It has some advantages in that it is very strong, and will bear greater weights when roof maintenance has to be carried out. It is very durable and will last indefinitely if protected against rust by means of aluminium paint or zinc spraying. It is, of course, somewhat expensive.

It has also some disadvantages, besides its expense, which directly reduce its popularity. It has a high conductivity, thus giving a high coefficient of expansión. This causes some difficulty when glazing as the metal expands and cracks the putty. Linseed oil putty is of no use and something which does not set hard but maintains a slight elasticity, such as Secomastic, should be used. This material has an added advantage in that it will adhere to the steel rafter without a preliminary coat of oil paint being necessary. An alternative is to use patent zinc clips to secure the glass.

GLASS

1. *Plate glass.* This is a very pure form of glass having no air bubbles and giving very little, if any, distortion of the light. One great advantage is that it has considerable mechanical strength, but its cost is very high which makes it rather prohibitive when used for glazing a large area.

2. *Sheet glass.* This is a poorer grade of glass but is most commonly used for horticultural purposes. It has a greenish tinge and tends to give excessive distortion, but experiments have proved that this does not unduly hinder plant growth. The standard size used is 18 in. × 20 in. and weighs 24 oz. per square foot. The better grades of sheet glass were largely imported from Belgium, which is famous for its high-grade sands used in glass production.

The disadvantage of ordinary glass, whether pure or otherwise, is that it will not allow all the ultra-violet light rays to pass through, but by incorporating certain metallic salts during manufacture it is possible to obtain a type of glass, known as 'Vita' glass, which allows the passage of ultra-violet rays right through to the plants. It is expensive and deteriorates after a few years. When this happens it gradually ceases to allow the passage of ultra-violet light rays and so becomes no better than ordinary glass.

3. *Double glazing.* The glass used in double-glazed greenhouses is essentially thin. This is necessary, firstly, so that a light framework can be used to carry it, and secondly to provide high light transmission. Because the glass is thin, heat passes through it easily, and the cost of fuel to maintain the necessary temperature in a house glazed with a single layer of this glass would be very high.

How can insulation be achieved without serious reduction of light transmission? Double glazing can supply an answer. It consists, instead of a single pane of glass, of two panes of glass spaced apart with a layer of still air, which transmits heat very slowly, confined between them. Using double glazing with an air space of ¾ in. reduces the heat loss to one half of that with single glazing. Since the heat is retained, less make-up heat is required, less fuel has to be burnt and so a smaller heating plant can be used.

There is thus a saving on both initial capital cost and running costs to set off against the increased cost of a double-glazed house as compared with a single-glazed one. Actually only two or three years are required to show a progressive gain. It is true that using two panes of glass does reduce the light transmission slightly (from 85% to about 75%), but for the domestic grower this is not of great significance. The double-glazed house used at Arkley Manor, my own home, measures 18 ft. × 8 ft., but the typical amateur's house measures 10 ft. 6 in. ×

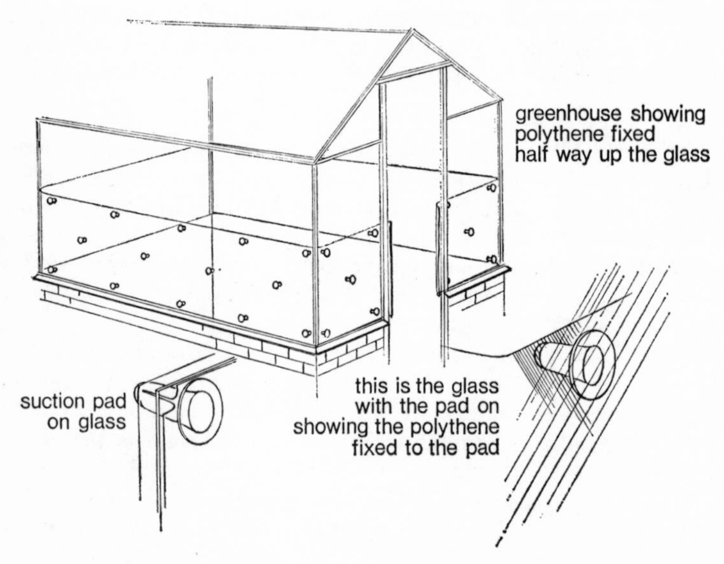

FIG. 1 How to double-glaze a metal greenhouse

8 ft. Those who have already erected their greenhouses may not see fit to double-glaze them – not an easy job in most cases. The obvious alternative to glass in this particular case is thin gauge polythene between 30 in. and 48 in. wide. Any metal or wooden-framed greenhouse can easily be double-glazed with this if special suction pads are used.

The gardener must start with two or three suction pads stuck to the inside of the window at the door jamb, and these hold the sheeting in place. The roll is then unwound, placing more suction pads on the glass at 2 ft. intervals. A second width of polythene can then be positioned in a similar manner above the first, and of course the bottom edge shares the existing pads. The warmglaze suction pads must be pressed firmly on to the glass (which should be freshly cleaned). The suction pads have stems approximately 1 in. long and ½ in. wide, and can be utilised in both wood and metal structures since the suction cup is simply pressed on to the glass of the greenhouse wherever support for the insulating film is required. The film is secured to the pad by a plastic dome-headed pin stuck into the top of the suction pad stem. Thus not only is the material held in position, but it is maintained at a suitable distance – about an inch – from the existing glass, giving the necessary air blanket.

4. *Substitutes for glass.* In the field of plastics there are three materials which may be used instead of glass:

a. Glass-fibre reinforced polyester, usually called GRP
b. Clear rigid PVC
c. Acrylic sheeting such as Transpex

All these materials have high strength properties, can be bought in large sheets, can be bent or moulded and are light in weight. They are all more expensive than glass, the Acrylic sheet being the least expensive at three or four times the cost of glass. Another disadvantage is that all three materials have a greater coefficient of expansion than glass, and this has to be taken into account by do-it-yourself experts.

The GRP, even when new, has a light transmission inferior to glass – much inferior. The plastic yellows with time and the surface accumulates dirt which is difficult to remove.

The PVC has better light transmission than GRP, but is still far inferior to glass. It discolours with age and becomes brittle at low temperatures.

The Acrylic sheeting has better light transmission than glass but is not resistant to scratching and tends to craze with age, resulting in considerable loss of transmission.

Glass undoubtedly holds its own – and Acrylic sheeting comes a poor second.

Polythene film has recently become widely used for greenhouse crop production. The houses are usually constructed as tunnels. The film has to be renewed every two years.

Construction

The positions of the outer walls should be marked out and trenches 9 in. to 1 ft. deep by 18 in. wide dug. Concrete footings are then placed at the bottom. These are made by mixing five or six parts of washed ballast and one of cement to the consistency of porridge, and placing a two or three inch layer in the trench. The brickwork is then built up from this foundation, using a plumb-line and spirit-level. The walls may be either 4½ in. or 9 in. wide, according to the size of house. High-grade bricks should be used to ensure that there is no crumbling due to frost, as may occur with those of poorer quality.

The woodwork is then built upon the brickwork, allowing sufficient space between rafters to fit the 18 × 20 in. or 24 in. × 24 in. glass panes. The sheets are laid in the grooves on the rafters which have already been puttied up with a good linseed oil putty, and then pressed down to ensure a watertight joint. The glass sheets are then held in position by brass brads.

When glazing it is necessary to allow about a quarter of an inch overlap on each sheet. This will ensure that when water drains down the roof it will not drip in between the glass on to the plants below.

The ventilators are fixed to the ridge by means of hinges, and may be opened and closed either individually or all together by the use of one of the patent gearing mechanisms. The latter method is considerably more labour-saving. There is also automatic ventilatory equipment for small greenhouses.

Wooden slat staging is usually constructed of 2 × ⅝ in. strips of a convenient length, nailed upon cross strips, allowing 2 in. spaces between each piece. The ideal material to use is some hard wood, and it is advisable to soak it in a strong copper sulphate or green Rentokil solution before erection. This will prevent it rotting and tends to destroy any unwanted diseases it may be harbouring.

The only objection to this particular type of staging is that when small pots are placed upon it, they must be put directly on to the slats or else they topple sideways into the intervening spaces. The spaces are necessary to allow the free circulation of air around the plants. Also a completely solid wood staging would be expensive.

Solid staging is usually constructed of corrugated iron or asbestos sheets upon which rough ashes or Lytag are spread, but this type of staging does impede free air circulation. Aluminium alloy strip staging is a possible alternative. It is possible to use special galvanised 8 or 10 guage wire netting which provides effective open type benching.

SHADING. Many plants grown under glass, especially flowers, need some protection against the hot rays of the sun during certain periods of their life. The reason for this, of course, is that the growing of plants under glass is not natural, and in fact the glass acts as a 'magnifier' to a certain extent and so increases the temperature and the power of the sun's rays. The effect is worse in small low houses.

The effect on certain plants is apparent by the scorching of the leaves and flower petals and the hardening of the young shoots, thus preventing their rapid growth. In flower crops the petals also fade badly. Various methods are used to prevent this by placing on or over the glass some kind of material which will reflect some or many of the sun's rays and, therefore, only allow a certain amount of sunlight through to the plants. The other extreme may unfortunately soon be reached by too heavy shading. The interior of the house is made dark and this causes the plants to become long and spindly, in which case they will not flower.

The gardener who works in glasshouses finds shading an added advantage as it reduces the transpiration rate of the plants, with the result that the amount of watering is cut down considerably.

The most common materials used for shading are lime and whitening. Both these materials are, of course, of a semi-permanent nature and cannot be removed easily – as soon as a period of dull weather occurs – without a lot of labour being involved.

Lime or whitening can be mixed with cold water in a convenient receptacle until a creamy consistency is reached. This mixture is then applied to the outside of the glass by means of a syringe or Solo sprayer until sufficient cover is given. If a *small* quantity of old oil is added to the mixture, enough to cause the liquid to emulsify, the whitewash will then stick to the glass in such a manner that rain will not wash it off easily. Cement should never be used, for once applied it is impossible to remove from the glass.

Removal of the shading is best done the hard way, i.e. by means of a coarse-bristled broom. It is possible to use a dilute solution of hydrofluoric acid in water. This is a *dangerous* acid because it tends to have a corrosive effect on glass. It should be applied with a soft brush on a long handle and must be hosed off immediately afterwards with plenty of water.

I find that the best solution to the problem is to have green plastic roller blinds inside the house – though automatic outside blinds are of course excellent too, where they can be afforded (see p. 42).

THE CEDAR-WOOD GREENHOUSE. Cedar-wood has become popular with manufacturers of greenhouses and with amateurs, as its rot-resisting qualities make it an ideal wood for outdoor use.

For any greenhouse it is essential to have the footings absolutely level, for in no other way can draughts and seeping-in of water at ground level be avoided, and this, of course, would be disastrous in winter. Cedar greenhouses are usually erected on a course of bricks, and this should be slightly smaller than the greenhouse in order to allow the walls to overlap the brick footings, thus facilitating drainage. Greenhouses should also be fitted with sills all round so as to drain away rain water and eliminate leaks.

If it is intended to do all the growing on staging, it is not advisable to buy a house with glass to ground level; such houses are colder in the winter than those with a solid base and there is no advantage in having daylight beneath the staging. The staging supplied by a firm like Bath's of Herne Hill, London SE24, is easily taken out and can be stored when not in use.

Since the use of cedar the days of normal greenhouse painting are over, and the cedar can be treated with a special preparation that repels water and maintains the natural colour of the timber. This is a wax-based waterproofing solution prepared to give long-lasting protection against water absorption. It is not preservative, for cedar is rot-resisting, but it does prevent rain from penetrating the wood and keeps the cedar in peak condition.

THE ALUMINIUM GREENHOUSE. One of the alternatives to a cedar-wood greenhouse is the aluminium alloy glasshouse, which of course needs no painting or maintenance – the first cost is the last! The alloy used is corrosion-resistant, and in the case of the Hartley, the alloy is also chemically treated and finished with stove enamel to provide a durable non-porous coating. Because all loads are transmitted to the ground equally, the house is strong and rigid, and there are no purlin posts to obstruct the work in the greenhouse.

Each pane of glass is enclosed on all four edges with a non-cracking and non-hardening plastic of indefinite life. This in turn is enclosed in a metal section, thus rendering the glass waterproof and draught-proof. No pane of glass is actually in contact with its neighbour; thus there is no over-lapping and no green algae can collect.

The erection of the Hartley or Clearspan aluminium house is simple. Gutters are built as an integral part of the structure, and there is no complicated internal bracing. The ventilators are balanced and pivoted, thus allowing cool air to flow in below the ventilator and the warm air to flow out above.

The doors are sliding – they run on plastic wheels and never rattle. The overhead runners are completely enclosed and thus exclude frost, snow and rain. There is a catch handle provided for locking purposes. An aluminium house is extremely light and thus it is only necessary to provide a single-brick base wall. This should normally be 2 ft. 6 in. high for a greenhouse, but need only be a few inches high in the case of the aluminium house, where the side glass goes down almost to the ground.

It is possible to buy aluminium alloy staging with asbestos trays for the plant houses, together with shelving which can slip on to the framework under the eaves. The heating of a Hartley or Clearspan aluminium house can be done in accordance with the instructions given in Chapter 2.

THE LEAN-TO. In addition to the plant house or tomato house, it is possible to have a lean-to or sun lounge. This may be of any length from 7 ft. 6½ in. long, but is always 7 ft. 3 in. wide and 7 ft. 9 in. high. The lean-to is erected on a two-course single brick wall.

CHAPTER 2

Heating the Greenhouse

TEMPERATURE has been called the most important controllable factor in the greenhouse. It markedly affects the growth, development, yield, and quality of all crops. During the year, and for some parts of every day, outside temperatures may be below optimum and controlled heat is therefore necessary for crops growing in a greenhouse. This may prove expensive, so accurate control is imperative. Bad temperature control is a waste of fuel – I have found that a 1°F error can increase fuel consumption by at least 9%.

There are two main requirements in greenhouse heating; maximum efficiency and no waste. In this way you get the maximum effect from the heat applied plus the right temperature. The choice of fuel is always a disputed question and must be decided by the individual after considering cost and availability. Solid fuel is still cheaper than oil in some districts, but many amateurs favour oil anyway because it saves labour.

The air in the greenhouse can be heated by air-heaters or by circulating hot water in a system of pipes. The pipework should be of small bore – for equivalent heat in a 2 in. pipe costs only about half that of a 4 in. pipe, but the volume of water is only about one-quarter. Such small-bore hot-water systems have an extremely low thermal inertia and are much more flexible than the older 4 in. systems.

All kinds of methods for the heating of greenhouses have been used in the past. Our great-grandfathers adopted a system of arranging for a flue to run either inside the back wall of a lean-to house or under the floor of a span house. Slow combustion stoves were used, and it was the hot smoke or air that did the actual heating. It was a dangerous system because if the lengthy flue leaked at any part, the smoke got into the house and caused trouble. The methods briefly described in the following pages are a good deal safer. (Automatic systems are described in the next chapter.)

Hot-water systems

Many greenhouses today are heated by hot water circulating in a system of iron pipes. The water is heated in a boiler by means of anthracite, coke or oil fuel and, as any boy studying physics at school will explain, the hot water rises to the highest point while the cold comes back to the base of the boiler to be re-heated. Hot-water engineers who install the pipes arrange for the correct rise of the flow pipes and the equal fall of the return pipes. It is sometimes necessary in large greenhouses or in a series of greenhouses to use what is known as a booster pump on the return pipes, so as to ensure that the water reaches the boiler regularly and evenly. An electric pump is probably best for this purpose.

There are two main types of water boilers in general use today:

The tubular boiler consists of a number of steel water pipes (2 in. in diameter) which are fitted into water chambers at each end by means of steel bracing rods. Rubber washers ensure a complete water seal. The whole boiler when erected looks like a rectangular box made of tubes. There is an opening at one end with special lugs on to which the boiler door fits.

The sectional boiler is of much simpler design and consists of a number of sections with hollow water jackets. These are joined together by means of bracing rods, and gaskets ensure that they are watertight.

With all boilers it is important to see that the damper and the ashpit doors fit perfectly, so that the draught can be controlled at will to give a bright fire quickly when needed or be closed down so that the fire burns very slowly. The fire-box door should always fit tightly to stop cold air from entering and cooling the unburnt gases and, more important, to prevent these gases being carried up the chimney and thus wasted before they have been burned.

The object of stoking should be to maintain an evenly burning fire, which should be slightly saucer-shaped, and thereby to ensure that all the gases produced by the fuel are burnt. It must be remembered too

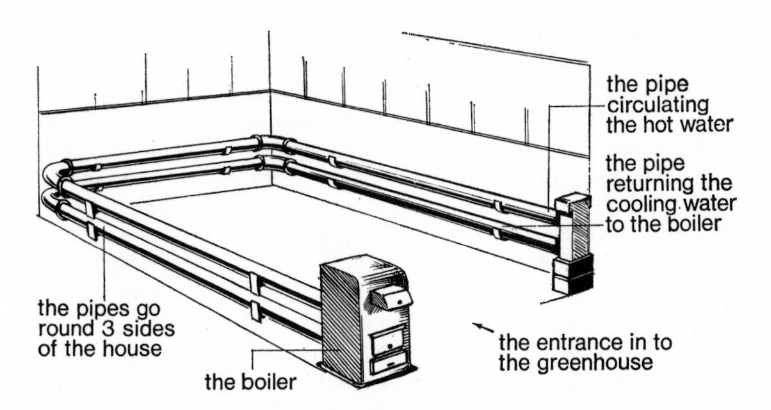

FIG. 2 The boiler and hot water heating system

that a hot fierce fire produces clinkers very quickly but a low fire produces only ashes.

The ideal fire depth will probably be from about 6 in., the object being to have a small fire which burns steadily, spread fairly evenly over the base of the fire-box. Once again fuel is often wasted by attempting to fill up the fire-box and thus having a depth of fire of, say, 1 ft. or more. It is quite possible – when the damper is closed – to use ashes over again, and so have a low fire of whatever fuel is being used, plus the ashes which will have a slight damping-down effect.

Most boilers are fitted with the correct number of flues, and these provide additional means for assuring that the water is heated in the most economical manner. It is essential that the flues are kept clean or the draught will be hindered. The damper, which can be closed or open, is usually fixed where the flue joins the chimney.

The main function of the chimney is to remove the waste products of combustion and to provide the necessary draught. In some designs the flue is arranged to enter the chimney at about 2 ft. above ground level, and this allows a space below for the soot and dirt to fall into without any blocking up taking place. It also prevents any rain water from draining into the fires.

THERMOSYPHON HEATING SYSTEM. This is the method most commonly used, adopting the principle that water rises when heated. The pipes must rise up through the house at a gradient of 1 in. in 10 ft., and the return pipes should have a similar fall back to the lowest point on the boiler which usually has to be below ground level. This method has the great advantage of being almost foolproof and is efficient if the piping is correctly maintained.

Pipes and water supply. All pipes should have a rise and fall of 1 in. in 10 ft., with a piece of ¼ in. gas piping at the highest point to allow air to escape as and when necessary. The pipes used are generally 2 or 4 in. in diameter, and 9 ft. in length. They are made of cast iron, the surface of which is left rough so as to give a greater surface for radiation. Each section of the piping has one end plain and the other in the form of a collar, which is made so as to fit over the plain end of the next pipe. The joints are sealed with a caulking material which should be cemented into position.

The boiler must be below the highest point of the piping layout to ensure efficient working of the thermosyphon system. This is, of course, not so important when other systems are in use. All flow pipes should be made easily accessible in case of bursts. Each boiler should also be fitted with a 'draw off' tap in order to enable the system to be emptied of water and refilled at least once every twelve months.

At approximately 5 ft. above its highest point, the heating system should be supplied from a water tank connected to the mains supply by means of a pipe and ball valve. Water should be allowed to flow from the mains supply until it fills the whole of the piping system and the tank, finally, of course, shutting itself off by means of the ball valve. When the water in the pipes expands upon heating the level rises in the tank, but not sufficiently to cause any overflow. If, on the other hand, water is lost by leakage or evaporation, the level will be readjusted immediately, automatically.

FUELS. There are three main types of fuel in use today: anthracite or hard coal, coke and oil. These can only be compared by knowing something about the British thermal unit, usually called B.T.U. This unit is the amount of heat required to raise one pound of water through one degree Centigrade. The table below shows quite clearly how the various fuels compare:

Fuel	*Volatile Gases*	*B.T.U. per cwt.*
Steam Coal	8–20%	6250
Anthracite	8%	7500
Furnace Coke	(Volatile gases driven off)	6300
Gas Coke	(Volatile gases driven off)	6350
Fuel Oil	—	9250

The size of fuel should, of course, be suited to the size of the boiler. It is no use having huge lumps of coal which will hardly go through the fire-box and will not begin to burn for a very long time.

Fuel oil is economical and, as can be seen from the above table, has a high heat value when burnt.

The simplest way of heating with fuel oil is to have an automatic burner such as the Phillips Oil Burner, which has been developed to provide the domestic grower with a simple, reliable, attention-free and economic method of heating a greenhouse.

The burner is a vaporising unit comprising a burner pan with two oil grooves, each of which is contained by two perforated metal shells. The perforations allow air to mix with the oil vapour, and continually provide the right air-oil mixture for any flame height and heat output within the limits of the burner. The top of the burner incorporates a deflector plate to ensure that the heat gases are diverted to scrub the walls and water passages of the boiler, ensuring the greatest heat transfer from radiation and also from scrub.

The oil level in the burner pan which controls the heat output is regulated by a float control. This is basically a needle valve metering-control which maintains its own constant head of oil. The constant head of oil in the control is important as this allows you to repeat the heat output at a control setting, as often as necessary and regardless of the amount of oil in the main storage tank. The control also provides a safety device which cuts off the oil supply to the burner if there should be any accidental rise in the oil level through maltreatment or dirt.

The heat output from the burner can be controlled by using two sizes of burner and four types of control. The small burner will provide a gross output of 20,000 B.T.U.s, say 14,000 B.T.U.s net, sufficient for a 70–80 ft., 3 in. diameter hot-water pipe at full heat; and the larger burner 40,000 B.T.U.s gross, 28,000 B.T.U.s net, sufficient for a 140–160 ft. pipe at full heat.

CONTROLS

1. *Manual* allows an infinitely variable position of the needle valve from pilot to fully open; a datum plate is marked with seven positions for guidance in repeating a setting.
2. *The water thermostat* will maintain a water temperature setting in the boiler system automatically, within the range of 140–180°F and within the capabilities of the system.
3. *The air thermostat* will maintain an air temperature setting for the heating system within the range 50–90°F and the capabilities of the system.
4. *The electric air thermostat* performs much as does the air thermostat, but the range will be greater, usually 32–90°F.

All thermostatic controls work on high and low flow principle. When the thermostat is calling for hot, the control and burner will be on high flame; when the thermostat temperature is reached, the control and burner will be on low flame. In use, it will be found that with thermostatic control the burner will balance heat production and loss and maintain a steady temperature without varying to extremes.

The oil burner therefore fulfils the main requirements of simplicity: a reliable, easy-to-install and inexpensive unit, a self-regulating burner, independent of outside power sources, providing heat for a minimum of three months – and usually a whole season – without attention, and a control which provides its own filter, safety factors, and precise control of the heat output. The economy derives from the fuel used, which is of constant quality and burns efficiently. It is delivered to a bulk tank and fed by gravity through large oil passages to the control and the burner.

PARAFFIN HEATER. If the greenhouse is considered too small for a coke boiler or one fired by fuel oil, a paraffin heater may be chosen. Gardeners do get satisfaction from a good paraffin heater, but it must be made expressly for greenhouse use. Each heater must have two essential parts:

the lamp which actually burns the paraffin and so produces the heat, and the section which actively distributes the heat.

Decide, as in the case of electricity (see p. 28), how much artificial heat is required in the greenhouse. Don't purchase the cheapest model. Remember that the price of the heater is actually related to the amount of heat it will produce. It is useless putting in a small heater and then bottling up what bit of heat there is by shutting the greenhouse tight. There is always a steady loss of heat from a greenhouse whenever the outdoor temperature is lower than the indoor temperature. Estimate how low the indoor temperature can be allowed to fall, and discover what is the normal lowest level to which the outdoor temperature will fall. If, for instance, the normal lowest temperature requirement is 40°F, and if you know that the outside temperature seldom drops below 24°F, then in a normal 10 × 8 ft. greenhouse you can say that for each degree of temperature difference the hourly loss will be 280 B.T.U.s. Since 40° minus 24° equals 16, then the rate of hourly heat loss is 4480 B.T.U.s. This then is the heat that the paraffin burner must be able to produce when turned right up. If it is less cold outdoors, the burner will naturally be turned down.

The heat given out by such a burner is in proportion to the rate of paraffin consumed. A burner consuming 1 gallon of paraffin in 32 hours gives out 4900 B.T.U.s per hour. A burner which consumes 1 gallon of paraffin in 64 hours gives half as much heat per hour. You can therefore calculate from these figures how much heat you can get from a burner before you buy it.

Not all this heat however is available in the greenhouse. A circulation of fresh air is just as important – a loss of 25 per cent of the heat generated must be expected – and this must be taken into account when choosing the heater. (You can expect to get about 4200 B.T.U.s from any large standard paraffin heater on the market.)

Do make sure that you choose the right type of heater. Those which conduct hot gases through a length of piping distribute the heat more evenly throughout the greenhouse than those in which the heat issues from one small hole.

The main danger of a paraffin heater is the fumes. Good heaters should be fumeless if properly looked after. They must be kept scrupulously clean, with the burner trimmed – preferably every day as directed by the makers – so that the flame is always level. This has to be done after cooling.

Always allow a little ventilation in order to provide the necessary oxygen for combustion. This reduces fume trouble and prevents the burner going out. Don't let the oil container run dry. Keep a check on the paraffin level, and always have a reserve supply in a can. To use a paraffin heater properly, one must be prepared to take time and trouble – which is why many have turned to electricity.

ELECTRICITY. This is a modern method of heating the greenhouse, and one which is becoming more and more popular. Apart from the running costs, it has every advantage – it can be thermostatically controlled, needs the absolute minimum of attention and takes up very little room. It is much used by amateurs, despite the expense of installation and the high cost of electricity per unit. There is also always the risk of the occasional power-cut, and if a break-down should occur on a frosty night, considerable damage to the plants can result.

For details of running costs, type and size of heater required, and automatic heating, see pp. 32–5.

HOW MUCH CURRENT IS NEEDED? Electricity is clean and easy to control and is therefore ideal for the busy amateur. The amount of heating equipment necessary is dependent to a certain extent on locality, height, and amount of exposure. A fair idea, however, can be obtained by using a formula produced by the Electrical Development Association, described below.

There are one or two factors that must be borne in mind.

1. The actual size or kilowatt rating of the installation will not affect the consumption. Thus the cost of running a thermostatically controlled installation may be kept low. Economy may reduce the first costs but it might result in insufficient heat to cope with a really cold night. Thus it always pays to err on the high side.

2. The heater will provide only the amount of heat shown on the label, i.e. 1 KW or 2 KW etc., no more and no less. The type of heater used, on the other hand, will affect both the proportion of radiated or convected heat, and the dryness of the heat, but the amount of heat will always be in accordance with the heater rating.

A gardener must know the usual difference expected between inside and outside temperatures. It is possible to maintain the greenhouse temperature against any outside temperature, but in order to keep capital cost low some sort of compromise must be reached. It is as well to assume that a minimum outside temperature of 20°F will be registered. Temperatures below this may be experienced, but they are usually of too short a duration to pull down the temperature inside the greenhouse to a dangerous level.

In order to discover the loading needed for the amateur's greenhouse, a simple formula has been evolved. This means working out what is called the 'equivalent glass area'. The area of the brickwork or wooden base should be divided by half and added to the area of the true glass. Metal bases have the same rate of heat loss as glass, therefore their total area must be added to the glass area. In the case of a lean-to greenhouse, the actual glass area must be taken plus half the wall area. In the case of a glass room or conservatory built on to a house where the next door room is occupied, work out the actual glass area and add to it one quarter only of the back wall area.

Having calculated the 'equivalent glass area', decide on the temperature you wish to maintain inside the greenhouse. Subtract from this the 20° that has been mentioned above.

The required loading in watts will thus be: (equivalent glass area in sq. ft.) × (required inside temperature − 20) × 0.37.

For example: if the equivalent glass area equals 216 sq. ft., and the temperature required is 40°F, you deduct 20 from 40, which gives you 20. Then you multiply 216 by 20 = 4320. This must be multiplied by 0.37, which gives you 1598.4 watts.

TEMPERATURE. Temperature is measured by means of a thermometer and for greenhouse work the Fahrenheit scale is generally used. In this scale, 32° is the freezing point of water and 212° its boiling point. The equivalent temperatures on the Centigrade scale are 0° for the freezing point of water and 100° the boiling point.

It may be necessary to convert Fahrenheit into Centigrade (and vice-versa), for many of the American horticultural publications use the Centigrade scale when describing temperatures. Remember $1°F = \frac{5}{9}°C$ and $1°C = \frac{9}{5}°F$.

To convert Fahrenheit to Centigrade or Celsine: Subtract 32 and multiply by $\frac{5}{9}$

For example: 59°F = ?°C
59 − 32 = 27 (59 is 27° above freezing point)
$27 \times \frac{5}{9} = 15$
therefore 59°F = 15°C

To convert Centigrade to Fahrenheit: multiply by $\frac{9}{5}$ and add 32

For example: 45°C = ?°F
$45 \times \frac{9}{5} = 81$
81 + 32 = 113
therefore 45°C = 113°F

(See p. 276 for temperature conversion table.)

HIGH SPEED GAS IN THE GREENHOUSE. The natural gas burning Shilton greenhouse heater is an entirely new concept in greenhouse heating. It is a natural gas appliance which brings (in fact) natural gas to the domestic greenhouse.

It is compact and flueless, automatic in operation, and simple to install. It incorporates a built-in, set-and-forget thermostat, which means that the amateur gardener need never again be caught out by a sudden frost. It can be relied upon to distribute clean even heat to all parts of the greenhouse.

Accurate thermostatic control on this Shilton heater eliminates fuel wastage and makes the heater economical in use. Natural gas is now a very cheap fuel, and with the Shilton every therm is used to full advantage.

The Shilton incorporates a flame failure device which prevents the main burners coming on unless the pilot is alight.

It also incorporates an additional benefit. By burning natural gas, the Shilton increases the carbon dioxide and moisture content of the greenhouse atmosphere, thus creating an environment which is beneficial to plant growth and yields. The principle of boosting the carbon dioxide content to provide an atmospheric fertiliser is widely accepted and the Shilton brings this advantage to the amateur.

Gas is piped to the heater by means of 8 mm. bore plastic-covered copper tubing; the plastic covering protects the copper, preventing corrosion from mineral salts in the soil.

Two gas taps, one on the outside wall of the house or garage and one in the greenhouse itself are needed; both taps have removable keys, and when locked at 'on' the gas supply cannot be turned off accidentally.

Measuring only 24 in. × 9 in. × 9 in., the Shilton will fit into the smallest space, and in fact it can be placed safely under shelves with only 3–4 in. clearance without damaging near-by plants. The heater is connected to flexible piping for portability.

METHODS OF RETAINING HEAT. It is curious, but true, that a wire netting fence 2 or 3 ft. away from the end of a greenhouse will break up any flow of cold air and cause a substantial reduction in loss of heat. A wooden fence, on the other hand, is often of no use as a windbreak unless it is tall *and* close to the greenhouse, for it only acts as a temporary deflector, as the wind will be deflected upwards and then down again on to the greenhouse. Trees are another very good protection against cold winds, but they must never be allowed to throw any shade on the glasshouse. Still air, of course, will not reduce greenhouse temperatures nearly so quickly as cold winds or air frosts.

The glass used in the house must be thin in order to transmit the maximum amount of light, but as a result its insulating properties are small. Any method for reducing heat loss from the glasshouse during the night is therefore welcomed by gardeners, as a considerable saving in fuel results.

Aluminium foil has been widely used for insulation purposes in recent years. Its highly polished surface reflects radiant heat, it is quite cheap to buy, it will retain a high polish even when exposed to the air, and it can be made in very thin sheets. It is manufactured in two forms: (1) sheets of paper faced on each side with very thin sheets of foil, (2) a flat sheet and a corrugated sheet of foil, stuck together in the same way as corrugated paper, so as to form air spaces. The latter is more efficient as an insulator, but is not so strong and is less easily handled than the former. The sheets should be placed in the evening inside the house within an inch or two of the glass.

PROTECTION AGAINST FROST. The subject of prevention of damage is naturally uppermost in the minds of all who grow plants in glass houses during the latter part of September and early October, for it is at this time of the year that the first frosts can be expected and gardeners begin to light their boiler fires or turn on the electricity.

Where frames are in use it is possible to cover them with tiffany or sacking in order to exclude the frost, or at any rate the bulk of it, and thus ensure no damage to the plants.

Very often gardeners will be caught unawares: the weather changes suddenly during the early autumn and a frost threatens, and the boiler fires have not been lit. In a case like this it is a mistake to shut the house down completely, for this causes all air circulation to cease and a damp stagnant atmosphere results which quickly cools down and, if the frost is severe, will cause frosting of the plants. It is always advisable to leave all the ventilators slightly open on the windward side, allowing a space of about 1 in. This will ensure gentle air circulation inside the house and so reduce the risk of frosting.

If the frost is very severe and the heating system cannot keep the temperature up sufficiently above freezing point, lighted candles inside pots at intervals down the house increase the temperature enough to keep the frost out.

When plants inside the houses do become slightly frosted it is often possible to save them by one of the following methods. The plants may be sprayed over with cold water, a few degrees above freezing point, and this is sufficient to thaw the frozen leaves and stems without damage. An alternative method, where the leaves are hairy, is to burn straw in the houses. The slightly warm smoke will thaw the plants and, providing they are not put in the sun, they will usually be all right, as long as the frosting has not been severe.

CHAPTER 3

Automation in the Greenhouse

THERE is no doubt that electricity can virtually take over the whole of the routine work of heating, ventilation, pest and disease control, and automatic watering. Further, the photo-electric cell in my own greenhouses takes over the control of the automatic shading.

Heating the Greenhouse

Having decided that the cleanest, 'least worry' method of heating your greenhouse is by electricity, you will naturally want some idea of the costs involved. You will need to take into consideration the running cost, temperature required, and size and type of heater.

RUNNING COSTS. Thermostatically controlled electricity has an advantage over other forms of heating, in that the current is used only when necessary, while temperature control is accurate and constant. To estimate the heating cost of keeping your greenhouse at 45°F, take the total equivalent glass area (see p. 28) and divide by the figure shown for your part of the country. The answer is, approximately, the average cost in pence per week for the seven colder months i.e. from October to April, with electricity costing, say, 1p. per unit.

Take a normal greenhouse with a brick wall side, 2 ft. high as in Fig. 3. If brick loses only half the heat of glass:

2 sides of glass 12 ft. by 3 ft.	=	72 sq. ft.
2 sides of brick 12 ft. by 2 ft. = 48 ÷ 2	=	24 ,, ,,
2 ends of glass, say, 8 ft. by 4 ft.	=	64 ,, ,,
2 ends of brick 8 ft. by 2 ft. = 32 ÷ 2	=	16 ,, ,,
2 roof sections of glass 12 ft. by 4 ft. 6 in.	=	108 ,, ,,
Thus the total equivalent glass area	=	284 sq. ft.

If you live in the warmer parts of England or Wales, you divide the total equivalent glass area by 13

If you live in the colder parts of the North and Scotland you divide by 6

And if you live in the rest of England, divide by 8

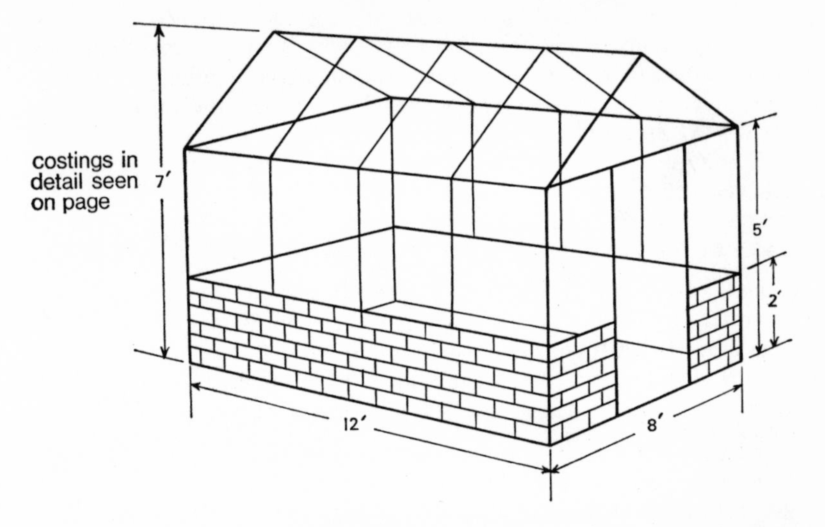

FIG. 3 Working out the electrical costings of heating a typical small greenhouse

This will give you the number of new pence it will cost in the winter to heat the house every week.
e.g. in London:

$$\frac{284}{8} = 35\tfrac{1}{2}\text{p.}$$

If on the other hand you are satisfied with having your greenhouse at 40°F. then you can reduce your costs by *half*, i.e. about 18p. a week.

TEMPERATURE REQUIRED. Many plants will survive in a temperature of 40°F, but it is always best to have heat in hand. For general purposes therefore you should think of a minimum temperature of 45°F. This temperature will keep most plants happy. For propagation or seed raising, you need 60–70°F. But then you need not raise the whole greenhouse temperature, but just install a small electrically heated frame.

SIZE OF HEATER. Now the size of the heater must be decided, remembering that you want to be able, when necessary, to raise the temperature in your greenhouse by, say, 25°F for the period when there is a serious frost. Aim always to have, as it were, a little heat in hand. So go back to the equivalent glass area, worked out as described, and multiply that figure by eleven. This gives the number of watts needed to raise the temperature of the house by 25°F. Using as an example the greenhouse illustrated on p. 33, the equivalent glass area is 284, and this multiplied by 11, gives 3124 watts. This in turn indicates that you need 52 ft. of tubular heaters.

Remember you can always discuss the details with your local Electricity Board.

TYPE OF HEATER

Tubular heaters were first used in greenhouses. They can be left unattended, under thermostatic control, for months on end. They will function efficiently for many years and are available in varying lengths or 'made to measure' to suit the needs of the house. Breakdowns are quite rare, and then only part of the heating apparatus seems to be affected.

Fan heaters on the other hand are popular because they provide a gentle movement of warm air, beneficial to plants. In spring and summer they will circulate air without applying heat, and they may be moved around in the greenhouse. The running costs are definitely higher, however, and I find they require more maintenance than other heaters.

The water-filled heaters, too, provide heat plus humidity. The hot water in the heater continues to dissipate heat after the thermostat has completely cut out and an even temperature level results. However, the water does need 'topping up' from time to time, and this can be a

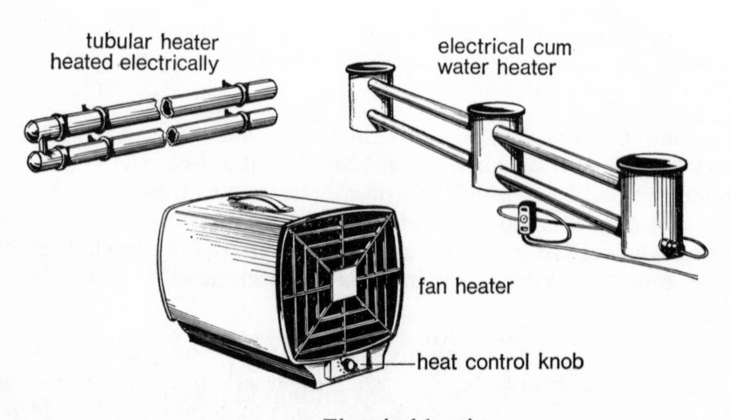

FIG. 4 Electrical heating

nuisance. There is for this purpose a shallow trough of water above the heater.

The choice of the heater is a matter of personal preference, but I prefer tubular heaters, which are well worth the extra trouble to install.

Warming the Soil

Where warmth is wanted for seed germination or for striking cuttings, or even for producing early salads, soil warming is a very efficient, economical method. Since the warmth is applied directly where it is needed, the waste can be negligible, and the current consumption, therefore, very low indeed.

Nothing is more simple to install: an insulated electrically heated wire is buried 5 or 6 in. below the soil level, and once positioned it requires no attention. The soil retains its temperature for a long time, so that when the heating is turned off, there is never a sudden drop in warmth. In a heated greenhouse it may be economic, for instance, to keep the air temperature at only 40°F, while an enclosed propagating frame equipped with warming wires provides any required extra bottom heat. Remember that, if necessary, you can have localised air heating as well. A propagating frame in my own electrically heated greenhouse costs only ½p. per square foot per week to heat.

There are two alternative methods available. The running costs are identical for both.

a. *The low voltage transformer.* This reduces the mains electricity supply to a safe voltage. The current is used to heat a plastic-covered galvanised iron warming wire laid under the sand or soil. Such wires are recommended where:

a. there is the danger of cultivations disturbing the warming wires (as in frames) and
b. for what is called "rotational cropping", where a warming wire is laid in each bed and then the one transformer is moved around for connection as necessary.

b. *The mains warming cable.* This offers a less expensive method and is widely used. It operates at full mains voltage, but is rendered completely safe by means of adequate insulation. If laid where cultivation by fork or trowel is anticipated, it must be buried 10 in. deep.

THE AUTOMATIC PROPAGATOR FOR GERMINATING SEEDS AND CUTTINGS. This 'greenhouse within a greenhouse' provides a nice warm bed on which to stand seedboxes and cuttings in pots, and maintains a soil temperature of 55–65°F without raising the temperature of the whole greenhouse. The automatic propagator consumes a mere 40 watts (my propagator costs only about 1½p. per week to run) and is regulated by a thermostat with a pilot light which glows when the thermostat is 'on'.

The propagator must be a self-contained unit and must be made completely safe because a low voltage warming wire is sandwiched between two layers of sand on which the pots and seedboxes may be placed. Sedge peat fills the spaces between and this acts as a heat insulator, conserving moisture and helping to maintain a uniform temperature in the bed. Full instructions for operating are supplied with the unit. The propagator should fit neatly on the greenhouse bench and should be movable. The overall dimensions can be 36 × 20 × 12 in.

Watering

MIST WATERING AND PROPAGATION. It is possible to water pot plants on the greenhouse bench by means of a mist which switches on and off automatically. The apparatus is simple. One arm carries a ceramic, the other a low-voltage contact. The ceramic absorbs moisture as and when the plants are sprayed; this extra moisture weight tips the arm, breaks the contact and switches off the mist spray. As the moisture gradually evaporates from leaves and the ceramic, the latter's reduced weight allows the opposite arm to drop, and the spray is switched on again.

In addition to watering plants, it has been established that cuttings root better and more quickly under controlled mist spraying. This method provides a thin film of water on all the leaves, which inhibits fungal growths, prevents wilting and allows softer and thus quicker-rooting cuttings to be used. Soil warming can maintain a high temperature within the rooting medium which, of course, should drain freely.

APPLYING A TRICKLE OF WATER. This method is used for greenhouse borders, benches, and pot plants – in fact, anywhere that requires a supply of water to be released every so often. A hose connected to the water tap slowly fills a plastic tank. When the tank is full, a float device releases 3 pints of water through a 23 ft. trickle line having 20 drip nozzles at 12 in. intervals (Fig. 5).

Each nozzle can be placed in the centre of a pot and all the pot plants in the greenhouse can be watered by this drip-drip-drip method. The water is fed carefully and regularly, and without any splashing or disturbing of the texture of the compost.

CAPILLARY METHOD. This system allows the pot plants to suck up the water they need automatically.

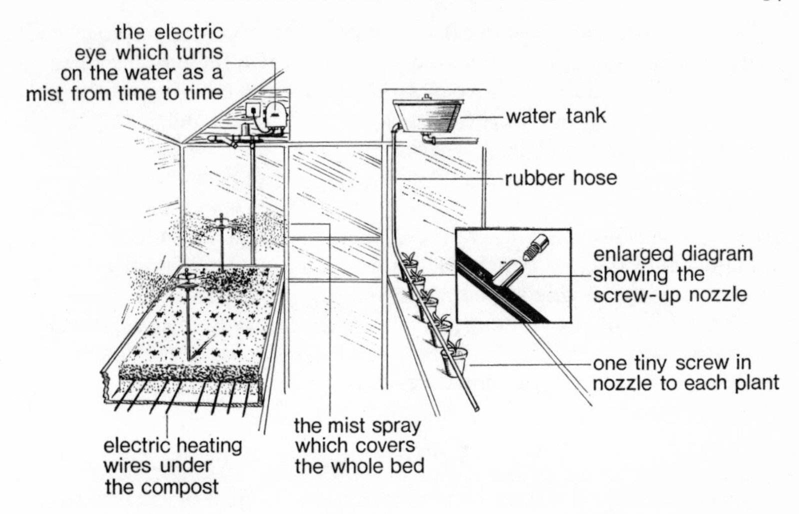

FIG. 5 Watering: *left*, overhead irrigation; *right*, trickle irrigation

The pots are stood on a 1 in. deep bed of sand which is kept moist by water drawn through a glass-fibre wick from an inverted bottle trough, and kept at a level 2 in. below the surface of the sand. The compost in the pots draws water in its turn from the sand, and remains in a moist healthy condition.

The system, of course, uses no electrical power and has no moving parts to wear out, so costs nothing to run.

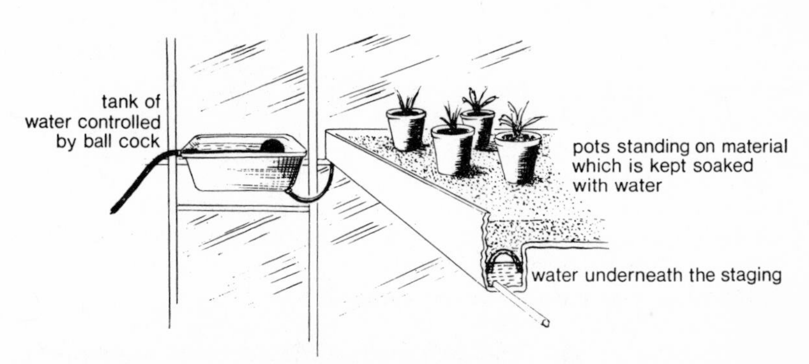

FIG. 6 Watering: the capillary method

When potting, no crocking is necessary. Clay pots will require small plugs of glass fibre to be inserted in the holes at their bases so as to ensure capillary contact between sandbed and the compost in the pot. Plastic pots are thinner at the base, so no such wick is necessary. Polypots are first-class and easy to clean.

Remember that the plants under this system are drawing just enough water from the soil at their roots, just as they would under normal growing conditions. A wide range of subjects can as a result be accommodated on the same bench. Bedding plants in plastic seed trays are happy growing on this system. You only have to firm the tray on the moist sand. Plants do undoubtedly thrive on this system and get all the humidity they need (Fig. 6). There is one disadvantage and that is that the soil is kept too wet in winter for some plants.

Ventilation

Excessive air temperatures can have an adverse effect on quality in the greenhouse, so it is important to limit high temperatures to as short a period as possible. With automatic ventilators sufficient heat can be removed to keep temperatures in the greenhouse within the right limits. If this can be achieved, the humidity too will be maintained at a reason-

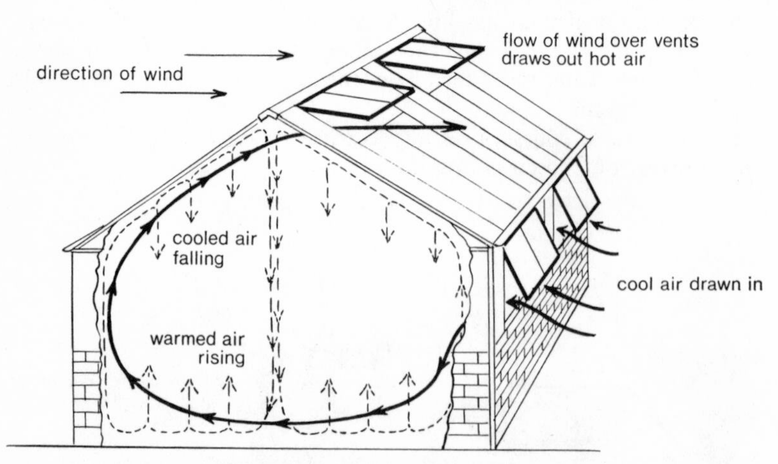

1. Dotted arrows show movement of air when all vents are closed & heater is working
2. Solid arrow shows flow of ventilating air

FIG. 7 A diagram showing what happens to the air when ventilating a greenhouse

ably low level. Where carbon dioxide is being given, there may be conflicting ventilation requirements (i.e. no ventilation may be advisable in order to conserve CO_2).

The common method of ventilation is of course the use of hinged ventilators. The official recommendation is for a ridge ventilator which should open at one-sixth of the base area plus as much side ventilator opening as possible. If it is impossible to provide bottom ventilation then a ridge ventilator must open in excess of one-sixth of the base area.

The trouble with the manual operation of ventilators is that it is impossible to achieve any good control of greenhouse air temperature. This usually results in excessively high temperatures in all greenhouses on sunny days. Hence it pays again and again to put in automatic ventilators.

AUTOMATIC VENTILATORS. These are entirely automatic and absolutely reliable. They ensure correct ventilation and cost nothing to run. There are several makes: I use one called the Ventmaster which comprises a bracket and special bar which consists of a narrow tungsten cylinder containing a mineral substance which expands or contracts at the smallest temperature change. It is this that operates the stainless steel

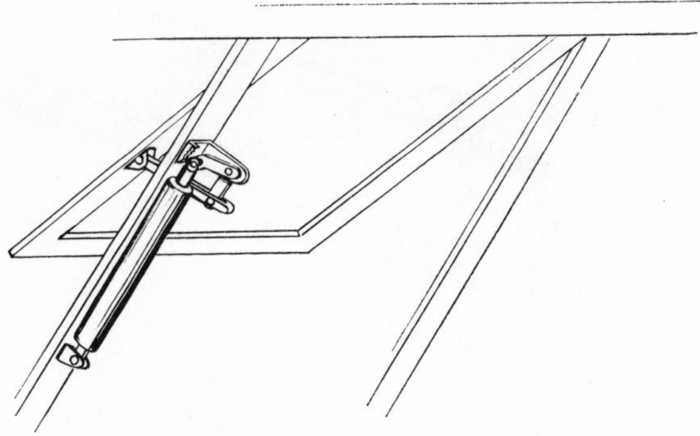

FIG. 8 Automatic ventilator

piston, which in its turn works a bell-crank level attached to a push rod, and this opens or shuts the ventilator. It also allows for wind strength and direction.

These gadgets can be attached to any ventilator in a few minutes with a screwdriver. The push-rod attachment will lift some 15 lb. and the actual temperature at which opening will occur is adjustable.

THE AUTOMATIC EXTRACTOR FAN. An alternative method of automatic ventilation consists of using an extractor fan arranged to draw air right through the glasshouse. The advantage of this system is that the rate of air movement through the crop can of course be higher than with natural ventilation. The disadvantages of fan ventilation are of course higher running costs and dependence on the electricity supply. There is also some difficulty in maintaining a uniform temperature all over the greenhouse when low ventilation rates are required.

Automatic Shading

Blinds that roll up and down automatically as the light intensity and temperature vary are wonderful time-savers. The reversible electric motor is controlled by a thermostat and a photo-electric cell. Thus there is a dual control of temperature and light. The blind rolls up as soon as either one of these elements falls below the predetermined level. Solar energy is utilised to the full when it is declining, because the photo-electric cell acts well at the end of a sunny day when the light intensity is falling and before the air temperature actually falls.

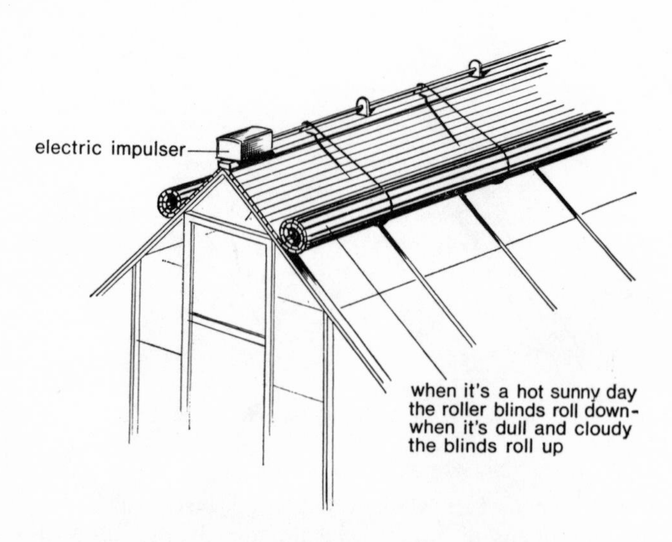

FIG. 9 Automatic rolling blinds

Partial Automatic Bulb Sterilisation

The sterilisation of bulbs and chrysanthemum stools are often a problem to the gardener, because equipment for such treatment is usually costly. A suitable small steriliser is made by Loheat Ltd. (Everland Road, Hungerford, Berks.), and supplied complete with two wire baskets, 18 in. long, 15 in. wide and 13½ in. deep. The steriliser is fitted with accurate thermostatic control which keeps the water at such a constant temperature that there is a variation of only 1°F (plus or minus ½°F) from that required. A forced water circulation is fitted, which is operated by a small electric water pump. The price, nearly £150, is rather high for most people – but worth it for the chrysanthemum expert.

CHAPTER 4

Greenhouse Equipment

IN addition to heating, ventilation, shading and even hygiene, you cannot run a greenhouse successfully and economically without the necessary apparatus. This may include anything from the humble bamboo to the very modern thermometer with the face of a clock, which you can read without actually going into the greenhouse itself.

I will therefore deal now with the equipment that may be necessary, and suggest sizes, types and even, in certain cases, makes.

WATERING CANS. Unfortunately there are numerous watering cans sold which are unwieldy and unbalanced to use. What is needed is a can with an easy-to-grip handle plus an easy-to-hold crossbar. This crossbar gives extra support and makes it possible for the gardener to hold the can in two hands if necessary so that the spout of the can can be correctly guided to individual pots.

It is necessary, therefore, to buy a can with a long spout. The first man to invent the correct type of can was a Mr Haw, and it is still possible to buy Haws Patent Cans – usually called Haws 'Pattern' Cans. They are perfectly balanced and can be quickly filled by dipping them into a tank or water butt, holding them by the long spouts.

One can water either with a rose or with the rose removed, as when feeding with diluted Marinure. It is often convenient with a small Haws can to put one's finger over the spout when feeding small pots and then to lift the finger a little to let out the necessary amount of liquid plant food.

There are various sizes of Haws cans: for large greenhouses use a 3 gallon can, plus a 1 gallon can for plants under shelves or for small specimens. For the moderate-sized greenhouse, a 2 gallon can, plus a small one, is ideal.

Recently a small plastic can with a special spout has been produced for using in a small greenhouse.

WATER TANK. It used to be thought a very good idea to have a water tank inside the greenhouse so that it could conserve the rainwater which collected on the roof, flowed into gutters and then into the down spout. It is certainly usually better than limey tapwater, especially as the latter not only causes spots on the leaves of the plant, but can create

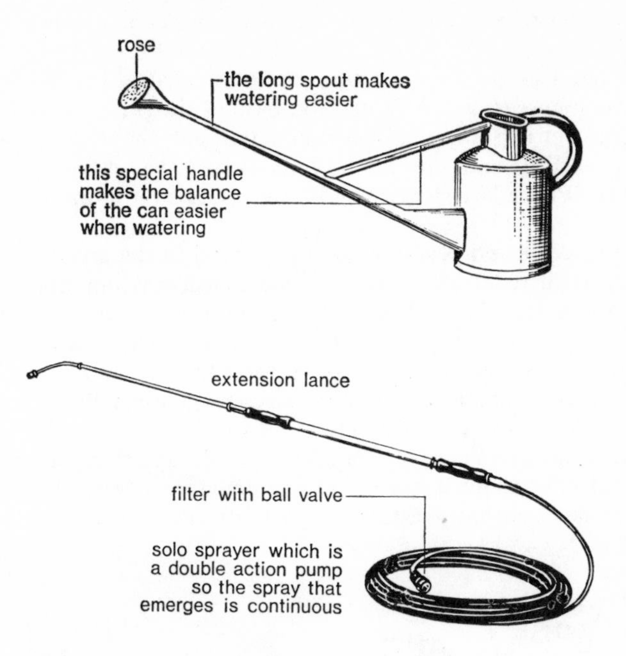

FIG. 10 Two methods of watering plants: *above*, a Haws watering can; *below*, a double action pump

difficulties in the case of specimens that definitely dislike lime. The tanks were usually sunk into the soil so that they were not in the way, but even so they did take up valuable space underneath the staging. The disadvantage of tanks in the greenhouse is that they can easily harbour diseases, and to use infected water could cause trouble. It is therefore better to have a tank outside the house, where the disease spores from the plants in the house are not likely to lodge. Those who do not wish to collect rainwater will have to arrange for it to be directed down a drain.

SYRINGES. Syringes are used for applying moisture to the leaves with a good deal of force, for damping down a house, for applying fungicides or insecticides, or even on occasions for watering baby seedlings. Syringes must be made of strong metal, so that they cannot easily dent. When this happens the leather plunger inside the syringe cannot move up and down properly. Syringes should always be kept spotlessly clean, and it is safer to have one for 'garden medicines', and another for clean water.

You can also get double action pumps shaped rather like syringes with a long hose (Fig. 10). When the base of the rubber hose attached to the sprayer is put into a bucket of water, and the upper part of the pump pushed up and down, a continuous spray issues from the nozzle provided. With a largish greenhouse this saves a lot of time.

POTS. For years and years the only pots used in the greenhouse were those made of reddish clay. It was said that it was important for the roots of plants that a pot should be able to breathe at its sides. Further, it was always claimed that the sides must be able to exude moisture. However, it has now been shown that these claims had no foundation, and that almost unbreakable plastic pots could give equally good results. Furthermore it has been discovered that small plastic pots need not be crocked because of a new type of drainage hole, and that plants in such pots need only about a quarter of the amount of water because the moisture does not evaporate through the sides. Thus these modern

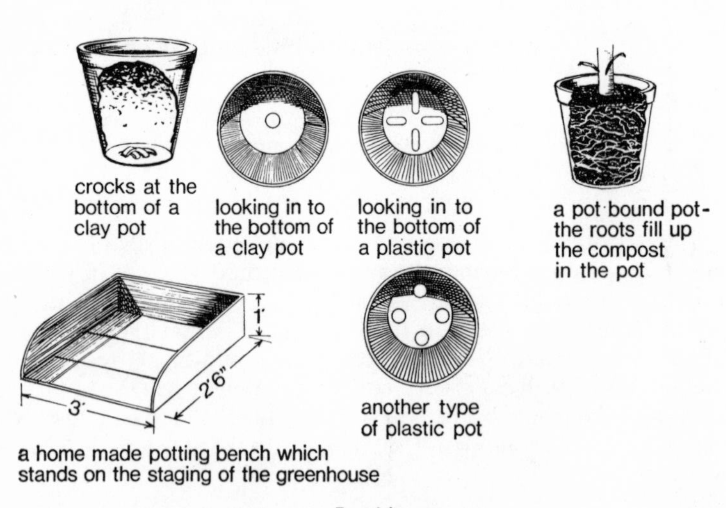

FIG. 11 Looking at pots

pots have become almost universally used. They are certainly lighter, and therefore easier to carry and to store, and they do not harbour diseases in the same way as clay pots.

Professional gardeners and shop assistants sometimes refer to pots by what seems a peculiar notation. They talk about 60s, or 48s – the 60s are the 3 in. pots, and the 48s are the 5 in. ones. The reason for this nomenclature is that you could make so many pots out of what was called a 'cast' of clay. A 'cast' had definite dimensions, and from this lump you could either make 60 3 in. pots, or 48 5 in. pots or 24 8 in. pots. Plastic pots do not have this particular classification as they are manufactured by a quite different process.

Plastic pots can be obtained in various colours – red, white, yellow, black and green – but many gardeners prefer to buy them in the typical clay colour which they are used to. The square pots save a considerable amount of space and are easier to feed with liquid fertilisers than round pots (as there is no waste over the sides). Furthermore, square pots have greater stability and the root space is slightly increased.

Plastic pots can be supplied with saucers, and if a little water is put into these the plant can absorb moisture through the base of the pot. Water should never be left for long periods in the saucer, since air circulation may be blocked.

POTTING BRUSHES. It is possible to buy circular brushes of various sizes which fit into the insides of different-sized flower pots. All you have to do is to twist the shaped brush inside the pot either with water if the pot is very dirty, or without water if the pot is fairly clean. Fig. 12 shows what the brushes look like. Those who have difficulty in obtaining them could write to the C.W.S. Brush Company, Wymondham, Norfolk, who will supply full particulars.

FIG. 12 Brushes for pots

DIBBERS AND POTTING STICKS. Many gardeners refuse to be bothered with dibbers, and they happily use any pencil they happen to have in their pocket instead. Some, however, specialise in chromium-plated dibbers which they use with great pride. They are about the thickness of a normal pencil and slightly pointed, and have the advantage of being easily cleaned. But some people claim that a metal dibber, because of its coldness and because it has never lived, is never as successful in the case of cuttings as a wooden dibber. There are also plastic dibbers.

A potting stick, or rammer as it is sometimes called, is just a piece of wood about 7 or 8 in. long of the thickness of a normal broom handle. The bottom part of the rammer should be wedge-shaped, and the top part should be rounded off. Normally the wedge-shaped part is used for pushing the compost down firmly all round the ball of soil at potting-on time, while the rounded end can be used to firm the soil in small pots in which seedlings may be pricked out.

SEED BOXES AND TRAYS. The seed box is normally made of wood, and its size is $14 \times 8\frac{1}{2} \times 2\frac{1}{2}$ in. Dipping the boxes in Rentokil green fluid makes them last far longer and also appear more unobtrusive. It is also possible to buy plastic seed trays of a similar size with drainage holes in the bottom. Some of these are a little fragile, so I prefer to use those which are firmer, even if a little dearer.

A SMALL PORTABLE POTTING BENCH. It is very nice to have a special potting shed in which the potting up and potting on can be done, but there are those who prefer to do the potting in the greenhouse. To avoid mess and waste they use a potting tray about 3 ft. wide and 3 ft. long, with the back and two sides 1 ft. high (Fig. 11). This is placed on the greenhouse bench and serves as a table.

PRESSERS. When sowing seeds in seed boxes, the compost has to be pressed down firmly, and it is as well to have a small board which will fit neatly inside the box for this purpose. With a small handle on the upper side of the presser, a smooth surface can easily be obtained before sowing and a certain amount of pressing down of the compost covering the seed can be done afterwards. Round ones for pots can also be bought or made.

SIEVES. The function of the sieve is to do the necessary separation of the larger particles of compost material from the smaller ones. When such composts as the Levington and Alexpeat are used, sieving is quite unnecessary as the compost is already sieved when it arrives.

When, however, the greenhouse owner likes to make up his own

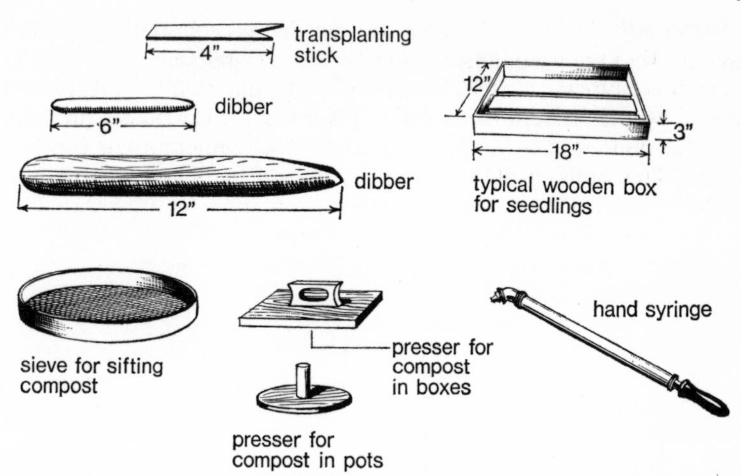

FIG. 13 Useful aids for greenhouse owners

compost under, for example, the John Innes formula (see p. 53), then it is necessary to sieve the loamy soil that is used and maybe even the silver sand, because in the latter case the idea is to use the coarser particles and not the very fine powder.

A fine wire gauze sieve, such as housewives buy for the kitchen, can be used and is very suitable when the instructions are to cover tiny seeds with a little sand. For the normal greenhouse, however a ¼ in. sieve should be sufficient to tackle any loam or garden soil likely to be used. John Innes Research Station specify a ⅜ in. sieve.

BAMBOOS. Because all supports used in the greenhouse should be as inconspicuous as possible, it is always a good thing to soak bamboos or canes in the Rentokil green preservative fluid. They will thus not only be less conspicuous but will last far longer.

TYING MATERIALS. The original material which gardeners always used was raffia. This was usually soaked in water for a minute or two after which it 'worked' more easily. (There is, incidentally, good and bad raffia.) The first alternative to raffia is a soft string called 'fillis' and one can buy it in either 3- or 4-ply, the latter being stronger than the former.

The second alternative is to buy special soft cotton twine, sometimes called twist or twill. This can be bought in reels and has the great advantage of being unobtrusive. The special balls of green twill are easy to keep in the pocket, and the 'string' doesn't get entangled like raffia.

THERMOMETERS. Unfortunately the normal greenhouse thermometer bought from the ironmonger is not always very reliable. I use the Clear View thermometer, a clock-faced type. The dial (which resembles the face of a clock) is kept in a suitable place outside the house, preferably near the door, while the 'bulb' recording the temperature is inside the house. The 'bulb' of the thermometer should be out of the direct rays of the sun.

PART TWO

Cultivation in the Greenhouse

CHAPTER 5

The Importance of the Right Compost

THE fact that we are now growing plants in the greenhouse without any soil at all would probably make some of the gardeners of the 1800s turn over in their graves. The history of this development, however, is quite interesting. Originally plants were grown in the greenhouse in soil mixtures which were supposed to be specially suited to the particular specimen. Gardeners experimented with mixtures of various ingredients and when one particular expert had produced first-class plants, he would perhaps disclose what mixture he had used. This was often recorded in apprentice gardeners' note-books, and in ancient 'tomes' written specially for the purpose.

The mixture, which could include soil, leaf-mould, rotted horse dung, bone-meal, farmyard manure, charcoal, dried blood, hoof and horn, and even burnt earth, came to be known as a compost. Gardeners tried hard to discover the medium in which the roots of the plants would grow best, and in certain books dozens of different composts are described in detail for various kinds of plants.

The first move towards standardisation took place in 1937–38. The John Innes Horticultural Institute experimented with a special new compost consisting of moderately heavy loam which had been prepared from stacked turves cut from a good pasture. This was mixed with coarse clean river sand and dust-free sedge peat. After the war, however, when it became extremely difficult to get the right kind of heavy loam to use for the compost, various new firms made up what they called a John Innes Compost, but using, instead of the correct loam, some type of sterilised garden soil. This resulted in failures and frequent complaints.

This led to the introduction by Michael Alexander of the Eclipse No-Soil composts, which would be sold in bags and whose standardisation could be guaranteed. These are now no longer available, but instead there are the Alexpeat composts. The Levington Research Station has also produced a peat compost called the Levington Compost.

Details of how to make up and use the three composts are given in the pages that follow. It will be noted that because of the difficulty in many districts in obtaining the right loam for the John Innes Compost, I have in almost all cases referred readers to the Alexpeat or Levington compost when dealing with the cultivation of different plants.

THE JOHN INNES COMPOSTS. These were originally designed by the

John Innes Research Station in the late 1930s, and depend for their success largely on the right use of 'medium clay loam' which unfortunately is not easy to obtain. It must be fibrous, should be of the sedge type, has to be clean and has to contain good grass. There are many so-called John Innes composts sold today which are not at all correct, because the loam used in them is poor or indifferent, resulting in an inadequate compost.

Most people living in towns or flats have neither time nor the facilities for making up potting mixtures, and they will buy John Innes composts already prepared. These composts are supplied in hundredweights or in bushel or half-bushel bags.

For those who wish to use the John Innes composts the requirements are as follows:

1. All the materials must be uniform in quality.
2. They must all be easily obtainable.
3. The loam must be as described above.
4. The sand must be river-washed and 'sharp' (60% of the particles must be between $\frac{1}{16}$ and $\frac{1}{8}$ of an inch in size).

Make up the John Innes Seed Compost as follows:

2 parts bulk medium loam (sterilised) sieved through a $\frac{3}{8}$ in. sieve
1 part bulk sedge peat
1 part bulk coarse silver sand (sterilised)
To each bushel* of the mixture add $1\frac{1}{2}$ oz. of superphosphate of lime and $\frac{3}{4}$ oz. ground limestone or ground chalk.

The ground chalk or ground limestone can be omitted in the case of plants like the baby azaleas and ericas which hate lime.

Never just guess quantities. Make certain that they are in accordance with the formula given. When sedge peat arrives dry, as it often does, wet it moderately before using it.

Mixing the Composts. Spread the sterilised loam out on the bench or floor in a layer 1 in. deep. Spread the sedge peat over the top evenly and the sand on top of that. Hold back some of the sand to mix with the fertilisers, as this makes it easier to apply them evenly. Use a trowel or small spade to turn over the three layers (loam, sand and sedge peat) until they are thoroughly mixed together.

Extra Food. The 'John Innes Potting Compost' is known as the J.I.P.1. It is the standard potting compost used for the majority of plants and must contain the special John Innes Base. Add twice the quantity of John Innes Base, and you make up what is known as the J.I.P.2. Add three times the quantity of John Innes Base, and J.I.P.3. is produced.

* A bushel of soil is contained in a box measuring 22 × 10 × 10 in.

Soil Composts. The essential requirements of a satisfactory soil compost are as follows:

1. It should be in good physical condition.
2. It should provide an adequate and balanced food supply at every stage of growth.
3. It should be free from all harmful organisms and substances.
4. Materials should be in good supply, and easy to obtain, reasonably cheap and uniform in quality.
5. Each material is for one purpose only.
6. Animal manure must not be used.
7. The method of making up the compost should be simple.

Loam. If loam is of the right quality, variations in peat and sand will be smoothed out. The test for good loam is that it should leave a greasy mark when smeared across paper. It should also be of a fibrous nature.

The best time for cutting turf for stacking is in early summer. A good turf stack should be 6 ft. high, 6–8 ft. wide, and any convenient length. The ideal size for turves is 12 × 9 × 4½ in. deep. The stack should be ready for use in six months. When breaking up the turf from the stack for use in potting composts the soil should be sliced to ensure it is evenly mixed.

Peat. Peat helps in aeration and regulates the moisture content of the soil. It contains 70–90% of organic matter, free from weed seeds, pests and disease, and therefore does not need to be sterilised. The fibrous granulated type should grade up to ⅜ in. The degree of acidity should be about pH 4 or 5.

There are two main types used in horticulture: sedge peat and sphagnum peat. Sphagnum peat is derived from the mosses and sedge peat from the sedges and rushes, and the two are quite different both in composition and value. In the experiments we carried out over several years sedge peat always gave better results than sphagnum peat.

Taking one ton of each, the two peats compared as follows:

	Sedge peat	*Sphagnum peat*
available humus	224 lb.	45 lb.
nitrogen	140 lb.	56 lb.
additional humus	70%	15%
protein	high	little or none
acidity	pH 5	pH 4

This illustrates the relative value of the two peats, and shows the importance of buying the right one for the job.

Sand. This should be clean, sharp and of glacial origin. Its value lies in the fact that it helps the drainage of surplus moisture. Coarse sands are of greater use than fine ones, grading from ⅛ in. to dust. Here again sand is naturally sterile and so needs no form of sterilisation.

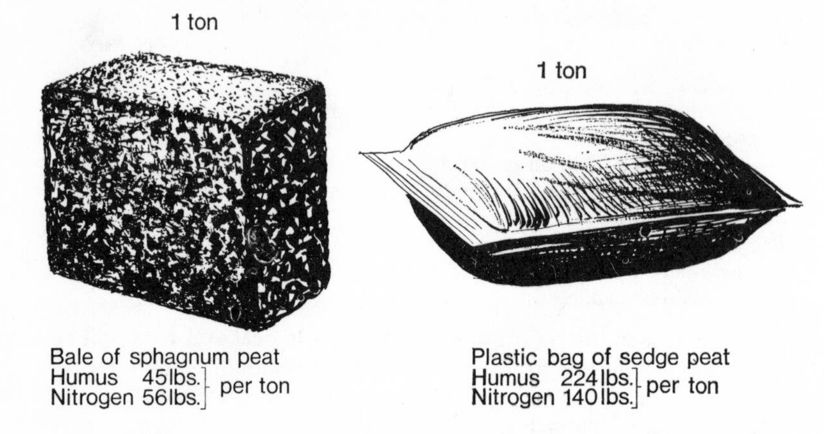

FIG. 14 Comparing sphagnum and sedge peat

TESTING SOIL FOR ACIDITY AND ALKALINITY. There are a number of simple tests to determine the degree of acidity or alkalinity of a soil, and one of the simplest is to take a sample of a soil and add a few drops of concentrated hydrochloric acid. If there is some effervescence lime is present and the soil is alkaline, but if there is no change the soil is either neutral or acid. This, of course, is only a very rough test.

The more scientific methods are based on a scale agreed by chemists and shown in degrees from 1 to 14. 1–6.9 shows an acid soil, 7.1–14 an alkaline one, 7 indicating a neutral soil. This is known as the pH or potential hydrogen-ion concentration scale.

The Comber Test involves the use of a reagent containing potassium thiocyanate or potassium sulphocyanide in alcohol. The reagent is colourless but turns various shades of pink or red when an acid or alkaline solution is added, and the deepness of the colour shows the amount of acidity: pink is alkaline, red is acid. A standard set of colour tubes or coloured card strips can be obtained which shows the actual degree on the pH scale.

The British Drug Houses Test relies on the fact that various dyestuffs change colour at certain degrees on the pH scale. Red, for instance, means acid, orange to yellow indicates that the soil is moderately acid, green shows it is neutral, while blue indicates alkalinity. The British Drug Houses manufacture a special colour chart showing these colours clearly and their equivalent number on the pH scale.

In all these tests except the first, the soil must be mixed with water before testing.

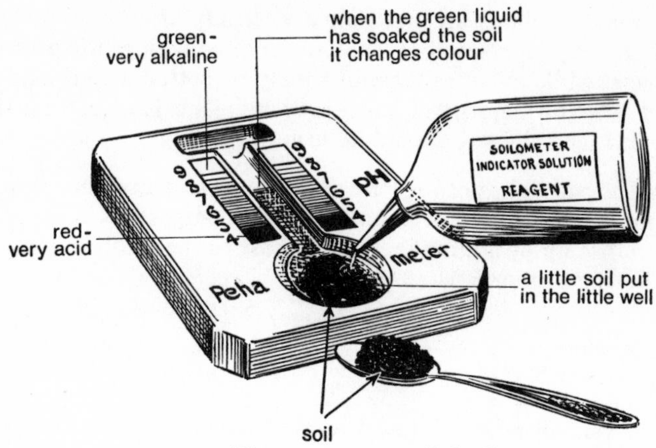

FIG. 15 The way to test soil for lime

SOIL STERILISATION. The term 'soil sterilisation' is rather a misnomer in this case, for if the complete process were carried out the soil would become totally sterile and would not support any form of plant life at all. What is really meant when this term is applied to soils is a method whereby partial sterilisation is carried out, destroying harmful bacteria and disease spores. The important and useful nitrifying and ammonia-producing bacteria are not affected owing to their harder outer covering.

The amateur can purchase an electric soil steriliser costing about £7. It is made of vitreous enamelled steel and holds half a cubic foot of soil. A 1500-watt immersion element produces the necessary steam which percolates through the soil until the required sterilising temperature is reached throughout the whole bulk. The size of this handy and effective unit is $9\frac{1}{4} \times 17 \times 9\frac{1}{2}$ in. The alternative is to buy loam that is already sterilised.

When using composts with partially sterilised soil it is also necessary, of course, to sterilise crocks and pots unless they happen to be new. It is also recommended that new clay pots be soaked before use, because if they are used dry they will absorb so much moisture that the ball of soil will dry out in a few hours. Crocks should be laid concave side downwards and it is better not to use rims or bottoms of pots. Crocks are unnecessary for plastic pots.

A uniform soil texture is essential to enable water to penetrate evenly and to allow sturdy growth. There should never be varying degrees of firmness and looseness in the compost in a pot. Pot therefore evenly and only fairly firmly.

Soft-wooded plants are not potted as firmly as hard-wooded subjects. For instance, cucumbers should be potted with gentle pressure of the fingers, while chrysanthemums should be potted with the help of a rammer. For the majority of quick-growing plants hard potting should be avoided, and the soil should be firm but springy to the touch.

Potting Compost. The potting base gives the correct nitrogen, phosphate and potash ratios for the compost. Potting composts Nos. 1, 2 and 3 may be made up by adding the potting base to the seedling compost in the following proportions:

	Potting base per bushel	*Potting base per cwt.*
Potting compost No. 1	11 oz.	$1\frac{3}{4}$ lb.
Potting compost No. 2	22 oz.	$3\frac{1}{2}$ lb.
Potting compost No. 3	33 oz.	$5\frac{1}{4}$ lb.

N.B. For small quantities: 1 bushel equals 8 gallons, and 3 oz. potting base is approximately 1 level teacup.

1. Add the potting base to make up the potting composts No. 1, 2 or 3, as required.
2. Mix very thoroughly.
3. Add water at the rate of 10 pints per bushel.
4. Mix and leave for several hours.
5. Fill pots evenly, level, and plant with a minimum of firming.
6. Water in and water again in two days during bright weather. In dull weather avoid over-watering in the early stages.
7. Keep the compost open in the early stages by watering gently. The compost must be uniformly moist.

The potting compost No. 1 may be used for seedlings which are likely to remain in the seed boxes for some time. It should always be kept moist, but not water-logged; water should be applied as a light fine mist in the early stages and subsequently with a fine rose.

The following are the recommended distances of the soil below the rim of the pot to allow for watering space: $\frac{3}{4}$–1 in. in 8 and 10 in. pots; $\frac{1}{2}$–$\frac{3}{4}$ in. in 6 in. pots; $\frac{1}{2}$ in. for 3 and 4 in. pots; $\frac{1}{4}$ in. for thumbs and tots.

The plant should be in active growth at the time of potting and the ball should not be dry.

THE ALEXPEAT COMPOST, formerly called No-Soil Compost. The Alexpeat Company has available a series of advanced composts developed from the no-soil idea all based on sedge peat. These are cheaper and are available ready mixed in a series of formulations e.g. Seed Sowing, No. 1 Low Nutrient, No. 2 All purpose, No. 3 Tomato, No. 4 Chrysanthemum and No. 2 Lime Free. All are mixed ready to use and store well.

Alexpeat Seed Sowing Compost is of course for seed sowing. The seedlings being pricked out *as soon as they can be handled*. If the seedlings have to be left in the compost for any reason they should be liquid fed. Under low light conditions No. 1 compost should be used for plants that dislike too much fertiliser e.g. Lettuce, Schizanthus, Antirrhinums. The No. 3 Tomato Compost is rich in slow acting plant foods and should be used under such conditions for plants that like plenty of food e.g. early bedding plants. For later plantings, however, use the No. 2 which releases its plant food more quickly. Strong growing plants like Chrysanthemums and Dahlias will need No. 4 later in the season.

Striking Cuttings. In the winter use the Seed Sowing Compost with sand. In the spring and summer, however, and particularly with strong growing plants such as Chrysanthemums, Dahlias and Pelargoniums, No. 2 Compost can be used, sand not being necessary.

Handling Alexpeat Composts. All the Alexpeat composts hold and retain a large volume of water when moistened – 2 or 3 times as much as John Innes Compost. When the compost dries out the water must be replaced, under warm conditions 2 or 3 times as much being necessary. Care should be taken not to overwater during the winter. The composts should *not* be firmed as watering will do this for you.

In Bags. The latest development in the use of this type of compost is in the growing of plants in a plastic bag laid flat on the ground. The Tom-Bag was the first of these used but it is now possible to grow crops of cucumbers, green and red peppers, melons, aubergines, courgettes, zucchini and ocra, in similar bags laid on the floor of the greenhouse.

THE LEVINGTON COMPOST. Amateurs who were given a trial sample of Levington Compost a year in advance of the official launching of the product in 1967 were the first to realise the completely different technique required for Levington compared with John Innes composts. The doubtful properties of many loams which were used in John Innes composts provided the biggest argument against them, although really good loams and properly formulated composts did give first-class results. Soil-less composts were introduced to overcome the doubtful loam factor, but it was the University of California's peat/sand composts which made things more difficult by demanding up to thirty different grades to meet the requirements of a complete range of plants. Variable results with this U.C. compost, as it was called, were experienced in Britain, due to different light, water, and other conditions in this country.

The Eclipse Peat No-Soil was the first loamless compost developed in Britain, which is now recognised as the Alexpeat Compost.

The Levington compost contains no sand or soil. Assuming that the properties of the peat used remain constant – and full research facilities are available to ensure that it does – gardeners rely on their growing skill to produce top-quality plants without having to worry about compensating any variables in the compost. Bedding plants do exceptionally well in it and antirrhinums are a complete success. Stocks germinate better in Levington than in John Innes, and I have found that chrysanthemums and dahlias are also perfectly happy in the Levington compost.

There are two types of Levington compost – Seedling and Potting. The main points to watch are to avoid over-firming and to ensure thorough watering. If seeds are to be covered, the smallest sieve used should be ¼ in., and light firming with a board after covering is all that is required. Thorough watering should be done before the seed is sown, and if boxes are covered with glass or polythene the compost should remain sufficiently moist until seed germination. Seed must be sown thinly to avoid root tangling and difficulty at pricking out, which should take place as early as possible.

Water thoroughly before pricking out into boxes, in order to stabilise the compost, but when pricking out into pots, do not water until after the seedlings are pricked out, and then give two waterings within 24 hours.

Sowing Seed in No-Soil Compost. Before sowing seeds in Alexpeat Compost, fill the boxes with the compost, level and lightly firm with a presser board, using just sufficient pressure to obtain an even surface. Water the box thoroughly, using a fine rose, ensuring that the compost is really wet. This soaking of the compost saves watering at the critical time following germination. Sow the seeds. Cover if necessary with a little compost. Cover with glass and newspaper in order to cut down the temperature, and to keep the moisture in the compost for quick germination. Immediately the seeds have germinated, remove the glass and paper and place the box in the light. If the box is in a frame or greenhouse with good humidity then the compost should not require any more water. If the seedlings do require watering, use a mist spray, which will not knock the seedlings down.

Root development of seedlings in No-Soil compost is faster than in John Innes compost. The seedlings must not be left in the boxes for any large root development, or root damage will occur on transplanting.

When cuttings are struck, the same watering technique is employed, except that the cuttings are left in the compost longer so as to obtain as much root as possible before transference. Water may be necessary whilst still in the cutting boxes and should be applied through a fine rose. Avoid overwet conditions if plastic seed trays are used.

Potting. Potting on should be loose – no hard ramming – and no crocks should be used. Pot with the very minimum of firming, and settle compost round roots with occasional waterings. The aim should be to keep the compost open and encourage quick root ramification; subsequent watering can be fairly generous, but avoid a peat-bog condition, especially in winter, and particularly with smaller plants in pots. Cyclamen must not be over-watered in the early stages after initial establishment.

It is natural for plants in No-Soil compost to grow fast at first, and they will need adequate space in which to develop. Later they will steady up and grow into normal plants. Plants in Alexpeat in clay pots flower earlier than those in plastic pots – in fact, there may be a swing back to clay pots for more expensive and difficult pot plants. Pot chrysanthemums in Levington grow softer than in John Innes compost, and because of this produce better and more even breaks, and ultimately a larger leaf, thus providing a well-furnished plant with top-size blooms.

Watering. When the seedlings are pricked out into boxes, test whether the compost feels dry by plunging your finger into it; if it does, water. Just feeling the surface of the compost is not sufficient. With pot plants, particularly those which are rather sensitive to too much water, e.g. cyclamen and cinerarias, there is only one way to make certain about the moisture in the compost and that is to remove the plant root ball from the pot. Invert the pot gently, holding the stem of the plant between the fingers and supporting the compost, tap the top edge of the pot smartly on the corner of a bench in order to loosen the root ball from the pot. Then test the root ball for moisture; if the compost feels drier than when it was taken from the bag it is safe to water. Try not to do this unless absolutely necessary, as plants naturally do not like being disturbed. One soon learns by experience the water requirement of the compost from the weight of the plant and pot together. There will be some amateur gardeners of the old school who will try pot-tapping in order to discover whether the pot is dry or not, but unless he or she possesses the ear of a 'railway carriage wheel tapper', this system is not recommended. The rule therefore is 'if in doubt – knock it out!'

If water is required, it can be applied from the top or from below. The common system of applying water to the top, by filling the pot up to the rim, is generally the safest. For most subjects the best procedure is probably to give a good watering and then leave, rather than little and often, but do ensure that adequate water has been given to moisten compost to the base of the pot.

A No-Soil Compost will 'take' more water at one time than John Innes compost, but the latter will hold it longer.

For composts for cacti and succulents see p. 200.

CHAPTER 6

Propagation in the Greenhouse

THE first priority for anyone considering propagation on any scale is absolute cleanliness. Some amateur gardeners make a practice of painting the inside rafters of propagating houses pure white every year. They do this for three reasons: to ensure that no disease spores are carried in the wood, to preserve the wood owing to excessive humidity maintained in propagating work, and to ensure the maximum reflected light.

Before beginning work the house together with all fixtures such as staging should be sprayed inside and out with a disinfectant such as cresylic acid, if there are no crops in the greenhouse at the time. The formula here is ⅛ pint to 5 pints of water. Low houses are ideal for propagation purposes, and for work early in the year those running east and west are the best.

SEED. Clean seed is of extreme importance as a number of diseases can be carried on the seed coats. It is, therefore, necessary to make sure that the source of seed is hygienic and reliable.

Seed should have at least a 40% germination rate and should be from a good 'strain', this factor being of greater importance than variety. Generally speaking the larger seeds in a batch will produce the better, healthier plants.

Time of sowing. This still largely depends upon the judgment of the gardener who aims at obtaining a mature plant for a given period.

Rate of seeding. Most amateurs sow far too thickly, and as a result hundreds of young seedlings germinate which are either too leggy or which damp off through over-crowding.

Depth of sowing. Sowing too deeply, in particular with small seeds, is fatal because the germinating seeds, having used up all the food supply they contained, die before the plumule has a chance of reaching the surface of the soil. A good general rule is to aim to sow a seed to a depth three times its own width – and that means sowing tiny seeds almost on the surface.

COMPOSTS. Seeds are sown either in boxes or pots, according to the number of plants required, and in the Alexpeat or Levington seed compost (see pp. 57–9). In the case of quick-germinating seeds it is hardly necessary to put a sheet of glass and piece of paper over the boxes before the cotyledons appear, except to prevent excessive drying. In the case of seeds that germinate more slowly this use of glass and paper is advisable, but the glass should be removed and wiped over each morning and then replaced. If watering is really necessary before germination, it should be done from below by dipping the boxes very gradually in water and allowing this to seep up through the bottom. Watering with a rose-can is not satisfactory for it may wash small seeds out of the pots or boxes. Similarly if the dipping is done suddenly, the soil may be forced out of the boxes.

Watering young seedlings with tepid water has been found unnecessary, but it may be advisable for the water to be of a similar temperature to that of the house.

PRICKING OFF. The gardener uses the term 'pricking off' to mean the transference of a tiny seedling from the box or pot in which it was sown into another pot or box. It is said that the term 'pricking' is used because the roots of the seedling are so small that the hole to take them can be made with a pin.

Most people allow young plants to grow too large before they transfer them, probably because they are frightened to handle really tiny specimens. With most plants one should prick off when the seedling has grown its first true leaf – sometimes called its first rough leaf. The first leaves to appear are two little round seed leaves, called cotyledons. Following these thickish cotyledons a different type of leaf appears which, though small, resembles the shape of the leaves of the adult

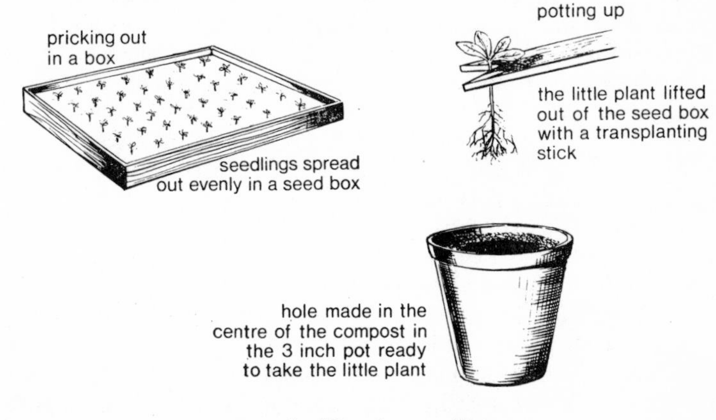

FIG. 16 Treating seedlings

plant. It is at this stage when the first rough or true leaf is seen that the pricking off should be done. After pricking off the box should be dipped very slowly into water until the moisture is seen to seep to the top of the compost. Pricking out at cotyledon stage is invariably best.

CUTTINGS. Many different plants are propagated by cuttings. When taking cuttings, remember that it is the young growths which give the best results as they root far more easily than old or harder portions. Such cuttings can be too immature or soft after a wet summer. This is even more true, perhaps, in the case of propagation by means of leaves. The cutting should generally have a reasonable amount of stored food within it, and therefore no material should be taken from weak or drawn plants.

Laterals or side-shoots are generally far more satisfactory for rooting than terminal shoots, the exception being the Spring tips of Autumn rooted plants like Geraniums. It is important to choose, if possible, material with a small pith area. Cuts are usually made at nodes or leaf joints, or just below them, but there are a few plants in which it is not so important, e.g. fuchsias, salvias and coleus.

Cuttings should not be allowed to flag before being inserted into seed boxes or frames filled with an Alexpeat compost. Expert gardeners make large postage-stamp sized cuttings (with part of a good vein). These are suitable for *Begonia rex* and butterfly-shaped cuttings of a similar size for streptocarpus. These are inserted upright in compost in rows. Shading should be given if the sun is very strong.

LEAF CUTTINGS. There are a number of plants, particularly ramondias, gloxinias, echeverias, begonias and streptocarpus, which are usually propagated by means of leaf cuttings because they all produce roots easily from their foliage. Well-developed leaves are chosen and hardly any preparation is necessary. I usually make shallow cuts in the veins of the leaf on the back side, especially just above where they join one another. The leaves are then laid, back downwards, on the propagating compost, which can consist of equal parts coarse silver sand and medium-grade sedge peat. Some gardeners have had better success with striking leaf cuttings in pure silver sand only.

I used at one time to peg the leaves down by using little wires shaped like hair-pins which could be pushed through the leaf just across a vein. Latterly however, I have had equally good results by laying the leaves on compost and just pressing them in a little.

With some plants, for instance streptocarpus, the top half of the leaf should be cut off to reduce transpiration. The leaf is then pushed into the compost upright so that it is buried to a depth of about $\frac{1}{4}$ in.

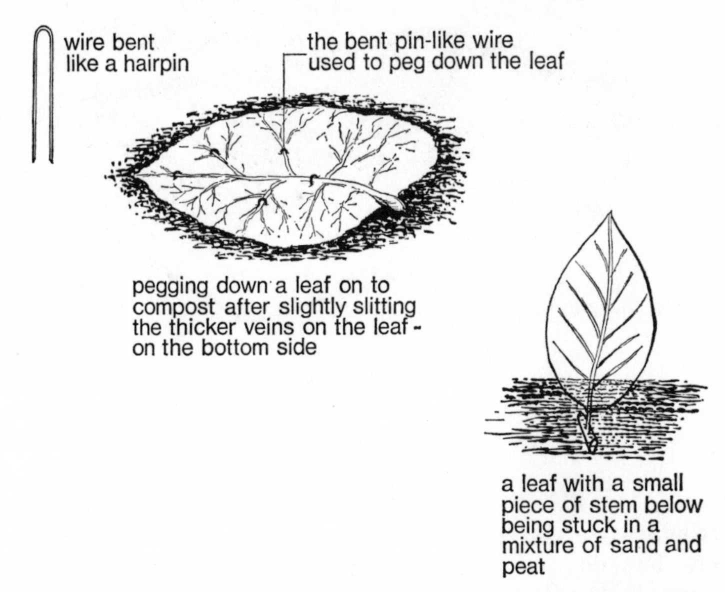

FIG. 17 Two methods of propagation by leaf cuttings

Leaf cuttings of gloxinias may also be taken by pushing them in upright, and the best results are achieved when these leaves are pushed in around the edge of a 6 in. pot filled with silver sand, or a mixture of silver sand and sedge peat.

On the whole, leaf cuttings strike best when the temperature of the greenhouse or propagating frame can be kept at between 60° and 65°F.

PROPAGATING FRAMES. A necessary adjunct for a greenhouse gardener is a propagating frame. This is nothing more than an ordinary frame made up on the greenhouse staging, with boards for the front and sides, the wall of the house for the back of the frame, and the top covered with a glass frame specially made to fit the width of the staging. Electrical bottom heat should be provided. The frame's chief use is to propagate cuttings and for this purpose it may be filled with a sedge peat and sand compost to a suitable depth, so that cuttings can be struck directly into it. The alternative is to put the boxes or pots of cuttings into the frames in which case a layer of peat 2 or 3 in. thick should be placed on the bottom of the frame to assist in retaining moisture and to help create that warm atmosphere necessary for successful rooting of cuttings.

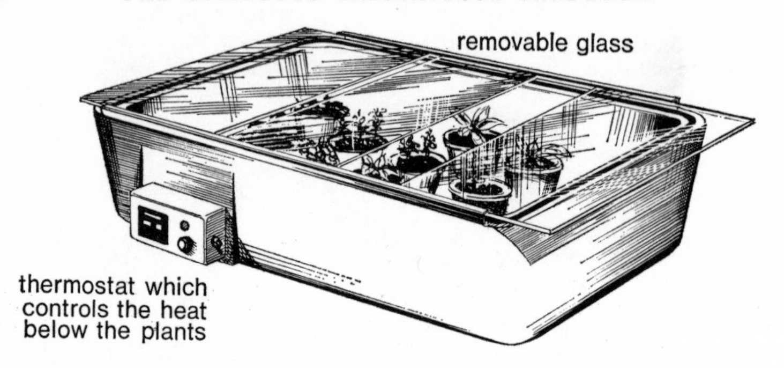

FIG. 18 A propagator

MIST PROPAGATION. The purpose of mist propagation is to raise the relative humidity inside the propagating frame, or even in the greenhouse itself, so reducing transpiration from the leaves. Cuttings used to have to be syringed frequently, but mist propagation means that the old-fashioned 'syringing over' is now done automatically.

Reference was made on p. 36 to the ceramic or electronic leaf, as some people call it. This is the device which controls the supply of water in the form of a very fine mist. Thus all through the day the mist is applied for half a minute or longer, not at regular intervals but whenever the ceramic or electronic leaf sends back an impulse which

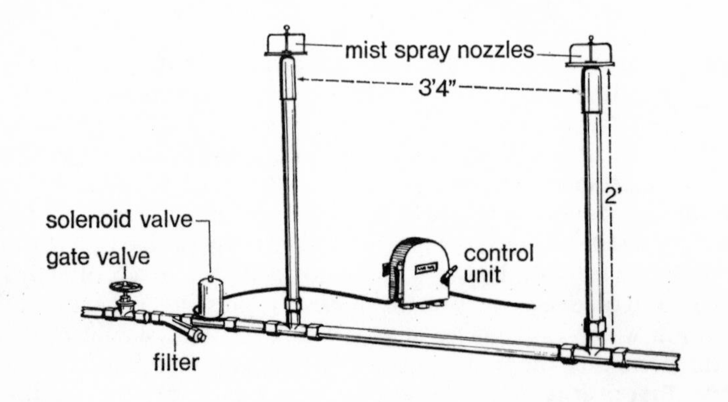

FIG. 19 Mist watering

indicates that humidity is too low. This method therefore keeps the leaves just damp, and at the same time ensures that the cutting bed is never saturated.

Those who want to install mist propagation in their greenhouses can get further information from Roberts Electrical Co. Ltd, Humex House, 11 High Road, Byfleet, Surrey.

USING HORMONES. Cuttings are usually taken just below a joint because the growth-promoting substances are concentrated there. Rapid rooting, vital to first-class propagation, depends on these substances, but there may not be enough present to cope with this type of propagation. Hormones, like other growth-substances, can be fed to plants.

Seradix is one of the most easily applied hormone root-forming preparations. It promotes vigorous rooting quickly, so that strong healthy plants are soon established. Different concentrations of hormone are required for different plant species. There are therefore powders formulated in three degrees of strength for treating soft, medium or hardwood cuttings.

Method of use

1. The base of the trimmed cutting is stirred in the hormone powder.
2. The surplus powder is removed by tapping against the rim of the tin.
3. The cutting is planted in the compost or sand.

A small container of the hormone powder will be sufficient for many hundreds of cuttings.

Examples

Use Seradix B (pink) for softwood and general cuttings:

African Violets
Carnations
Chrysanthemums
Coleus
Dianthus
Fuchsias
Hydrangeas
Ivys
Pelargoniums

Seradix B (white) for medium cuttings:

Briars
Cydonias
Viburnums
Clematis
Rubber plants

Seradix B (grey) for hardwood cuttings:

Azaleas
Camellias
Wistarias
Bougainvillaeas
Lilacs

CHAPTER 7

Potting On and Potting Off

WHERE necessary throughout the book, instructions are given as to times of potting on, methods of potting, and sizes of pots to use. This chapter is designed to deal with the whole question of pots, potting on, potting off, potting back, and so on.

CLEAN AND DIRTY POTS. It might seem unnecessary to mention the importance of having perfectly clean pots, but I have seen truly dirty ones used by amateurs again and again. To make certain that the ball of soil emerges in perfect condition from the pot when it is placed upside down and tapped slightly on a potting bench, one must start with a perfectly clean pot. Further, dirty pots can carry diseases and a really good washing helps greatly. Some people use a tablespoonful of liquid Derris and a tablespoonful of Captan in 2 gallons of warm water, and scrubbing brushes or the special pot brushes described on p. 45.

Plastic pots do not 'hold' the old soil or compost as strongly as clay ones, and their smooth non-porous surfaces tend not to hold diseases and are easy to clean. It is advisable to soak thoroughly new clay pots in clean water for about three hours before they are actually used.

CROCKING. In the case of the normal clay pot, there is a fairly large drainage hole in the middle of the base, and to prevent the compost from washing through, a broken piece of pot called a crock is put over the hole, concave side downwards. Over this it is advisable to place some coarse sedge peat, which acts as a 'buffer' between the compost above and the crock below, and ensures better drainage.

There are cases throughout the book where it is suggested that numbers of crocks should be used at the bottom of pots, especially for those plants that require perfectly drained, open compost. On the whole, though, it can be said that annuals require more crocking than perennials, and that plants that are going to be potted on require less crocks than plants that are going to grow in what may be called their final pots.

POTTING UP. This is a term usually applied when a little plant, a seedling, or a cutting is first potted into a small 2 or 3 in. pot. The compost is put into the pot over the drainage material until the pot is

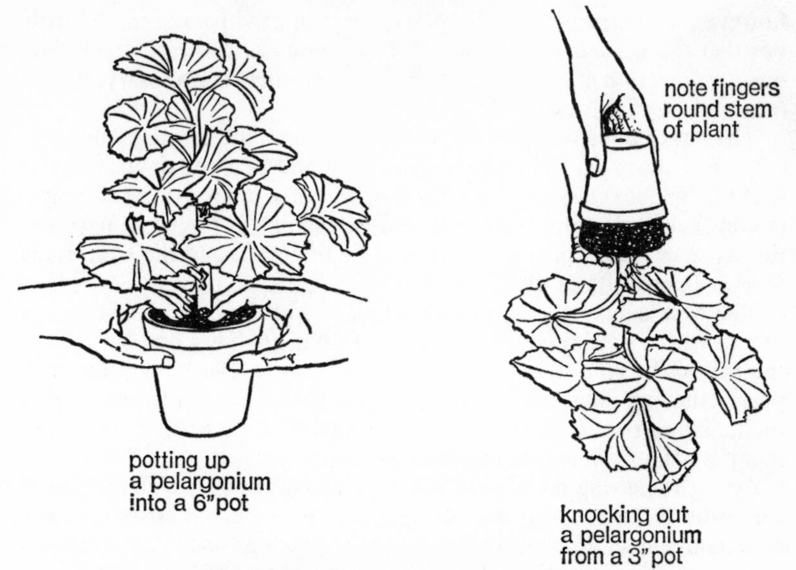

FIG. 20 Potting up a pelargonium

absolutely full. It is then pressed down, either with a clenched hand or a circular wooden presser shaped to fit exactly inside the pot, but there is never any need for the compost to be over-firm – after all, the normal soil in cultivated ground out-of-doors is never as hard as a rock!

On the whole, the John Innes compost needs to be a little firmer than the Alexpeat compost, and the Levington compost less firm again. In the latter case tapping the base of the pot on the bench two or three times will usually settle the compost as firmly as is necessary. When sufficiently pressed down and firmed, the compost level should be half an inch from the top of the rim of the pot. A hole is now made with a dibber, or even with a finger, big enough to receive the roots easily. The roots of the little plants are then inserted so that they are just a little below the level they were at when growing in the box or pot. Generally speaking, seedlings should be potted up as early as possible, i.e. as soon as it can be handled in the seed leaf stage, while rooted cuttings should be potted when the roots are about ½ in. long. If you leave the latter any longer, the roots are apt to become damaged during the potting up process.

POTTING ON. Plants are allowed to grow in their first pot as long as possible. The reason for this is that a plant in a 3 in. pot takes up far less room in the greenhouse than a plant in a 6 in. pot. It is important, however, to pot on long before the roots are so massed in the little pot that the gardener calls the plant 'pot bound'. Generally speaking, one pots on from a 3 in. pot to a 6 in. pot and then, if necessary, from a 6 in. pot to an 8 in. or 10 in. one.

The larger pot should be washed, crocked and filled with compost as described above for the smaller pots. It should be filled to such a level that the ball of soil from the previous pot, when stood on the compost, is ½ in. below the rim. Naturally, before the ball is placed in position, the crock or crocks at the base should be removed, and the basal roots may be opened out a little.

The new compost is then poured all round the ball of soil and is firmed with a potting stick as described on p. 46. The compost around the ball of the plant must always be made a little firmer than the compost in the first pot. If this is not done, the ball of soil tends to dry up as the moisture travels down the looser compost on the outside. So ensure that the ball of soil and roots are moist before potting.

After the potting on, a good watering should be given, the leaves of the plant syringed over, and the pot put back on the staging of the greenhouse at a temperature similar to the previous one. The exception to this, of course, is when the pots go into a frame for the time being.

Some people like to slightly alter the shape of the ball of soil from the first pot, squeezing it with both hands in an endeavour to make it roughly the shape of a rugby football. They claim this makes watering easier and more effective in the new pot, but I have found that it is much more likely to damage roots, and I condemn it. In any case, see that the potting soil is moderately dry at the time of potting: both the ball of soil and the compost should be neither too wet nor too dry.

POTTING BACK. It is sometimes necessary, at the end of the season or at some other time, to try to make the plant grow successfully in a smaller pot. To do this, you must knock the ball of soil out of the pot carefully, and then rub off what are called the 'shoulders' of the ball of soil, together with some soil all round the ball, so as to reduce the bulk of compost and fit it into a smaller pot. It is not an easy thing to do except towards the end of the season, in the case of perennials, nor is it usually advisable.

POTTING HINTS. It is better to firm the compost in small pots with the fingers. Most people like to use their thumbs, but the latter are apt to press too hard. The aim should be to pot so that there are no cavities in the compost and so that the top of the compost is absolutely level. If the top is not level, the water will always run to the lower part and the ball of soil will not be moistened evenly.

The operation of turning out the ball of soil from the pot is done as shown in the drawing. Two fingers of one hand are placed on either side of the base of the plant, and on the compost itself. The other hand grips the base of the pot, which is turned upside down and then given a sharp tap on the edge of the bench. The ball of soil should then come away cleanly from the pot and it is then that the crocks at the base of the ball are removed.

The only time that plants need potting really firmly is in the case of hard-wooded shrubs like azaleas, deutzias, lilacs and acacias, which are usually potted up during November or December with the idea of putting them in the greenhouse and forcing them into flower.

Another exception is the orchids, which are generally potted up into osmunda fibre plus sphagnum moss, in as small a pot as possible.

CHAPTER 8

Watering and Feeding

Watering

1. Plants should be watered sparingly:
a. on wet, dull and frosty days;
b. if the house and outside temperature is low;
c. if plants are in resting condition or foliage is dying down;
d. if the plants are naturally used to dry conditions;
e. when plants are slow-growing;
f. when plants have been newly potted.

2. Plants are watered freely:
a. on hot windy days, and in hot rooms;
b. when they are used to moist conditions;
c. when they are in active growth.

3. Methods of testing for dryness in pot plants:
a. tap the pots. If dry there is a clear-ringing note, if wet a dull-sounding one (this method is, of course, no good if the pot is cracked or when plastic pots are used);
b. soil colour is light if dry, dark if wet;
c. feel the compost;
d. the pot will be heavy if the soil is wet and light if it is dry.

4. Pots should always be placed in an upright position and the watering space filled right up.
5. When watering newly potted plants for the first time a rose should be used.
6. When roots are disturbed by transplanting from box to pot the soil should be sufficiently moist to enable the roots to penetrate the new compost before further watering is necessary. Always water in.
7. A dry plant should never be potted on and the soil ball should be as moist as the new compost. A golden rule to observe is to water a plant thoroughly when it needs it and then leave it alone. Do not just sprinkle plants at frequent intervals.

Syringing encourages growth and softens stems of plants which tend to become hard and wiry. It maintains a humid atmosphere and reduces transpiration in newly potted plants. It also checks pests such as red spider. The ideal is to use an actual syringe which gives a fine mist, but as commercially this type of syringe is impracticable a watering can with a rose is used instead or a trigger-controlled lance and rose at the end of a hosepipe.

Feeding

When considering feeding plants, the chief processes involved in plant growth and development must be taken into consideration. These are as follows:

1. The intake of water and minerals by the roots.
2. Photosynthesis, i.e. the building up of sugars by the plant from carbon dioxide absorbed by the leaves and the water which is taken in through the roots. This process is assisted by the green colouring matter called chlorophyll in the leaves and by the energy from the sun.
3. Respiration, which is the combination of oxygen with the sugars built up in photosynthesis resulting in the release of energy which is used for plant growth.
4. Transpiration, which is the loss of water mainly through the leaves.
5. The building up of the complex compounds, such as proteins, which form the living material of the plant.
6. The storage of products such as starch in the tissues of some plants, e.g. tubers, bulbs and corms.

No hard and fast rules can be laid down for feeding, since much depends on the environment. For example, plants grown in a warm house require different treatment from those raised under cooler conditions; plants in small pots require more feeding than those in large ones. Light, water supply and humidity also affect the nutritional requirements; more nitrogen is required where there is plenty of light, while in conditions of low light intensity more potash is needed. For this reason it is not advisable to give nitrogenous fertilisers in the autumn and winter.

PRINCIPAL ELEMENTS FOUND IN PLANTS. There are ten major elements found in plants and a small number of trace elements. Of the major elements carbon and oxygen are obtained from the air, while the remainder are obtained from the soil, these being hydrogen, nitrogen, phosphorus, sulphur, calcium, potassium, magnesium and iron. The trace elements are manganese, boron, copper, zinc and molybdenum. All are necessary for healthy plant growth.

In some plants there are other elements (sodium, chlorine, silicon and aluminium) which apparently play no part in growth.

Mineral elements have to be present in a special form before absorption can occur; for instance, nitrogen should be in the form of nitrates, phosphorus of phosphates, and sulphur of sulphates. Calcium, magnesium, potassium, copper and zinc are absorbed as sulphates, chlorides and nitrates, and iron is taken up in the form of either ferrous or ferric salts.

Calcium is found mainly in the leaves, and its main function is to neutralise organic acids formed by the process of respiration and especially in plants containing oxalic acid.

Magnesium is an essential element of the chlorophyll (the green colouring matter) molecule, and if lacking will indirectly cause yellowing of the leaves.

Nitrogen forms 40–50% of plant protoplasm. It is a major element in plant growth and a constituent of chlorophyll. Nitrogen causes lush green leafy growth.

Phosphorus is another essential element in the formation of protoplasm. It assists in root development and encourages the maturing of the plant.

Potassium is concerned with cell division and is an essential constituent of cell protoplasm. Its effect on the plant is opposite to that of nitrogen, for it encourages 'hard' growth.

Sulphur is an essential element found in the protoplasm.

Iron is another important factor in chlorophyll formation. It is a substance which slows down or speeds up chemical reactions without itself undergoing any change.

The most important elements as far as the plant is concerned are nitrogen, phosphorus and potassium, and these can be applied in the form of liquid manure or organic fertiliser.

Occasionally deficiencies occur in some of the minor elements such as boron, manganese or magnesium, and these have to be added, but if the soil is kept well supplied with organic matter this should rarely be necessary.

Humus. When the greenhouse borders are used, they must be properly prepared, and this usually means forking in plenty of organic matter in the form of well-rotted compost. It is essential to keep up the supply of humus.

Humus has been described as a complex residue of decayed animal and vegetable matter together with substances built up by the soil fungi and bacteria, and its importance in the soil can hardly be over-estimated. It is a jelly-like substance which prevents the soil from being just dust, it stores water and plant foods, and prevents valuable minerals from being washed away. It also helps to keep the soil open and friable, thus ensuring good aeration. A soil with plenty of humus present is said to be in good tilth.

Feeding Pot Plants

In the case of the various soil and no-soil composts, their composition is such that all the food required by the plant in its early stages is supplied in the compost. When the plants become well established, some additional food is often necessary and the easiest way to apply this is in liquid form.

There are several proprietary liquid manures on the market today, and the great advantage of buying one's manure in a bottle is that the exact composition is known to the grower, so he can tell just how much nitrogen, phosphate and potash his plants are receiving. Also, when applied in liquid form, the manure is distributed evenly throughout the ball of soil, each particle of soil becomes coated with a thin film of the nutrient solution, so that every root hair can absorb it and the plant foods become immediately available. When solid fertilisers are used, a certain amount of check to growth occurs because of the high concentration of the manure in some parts of the soil. A further advantage of liquid manuring is that the plant can be supplied with its food and drink at one and the same time, so saving a good deal of labour.

As has already been pointed out, environment plays a large part in determining the amount of food required by a plant, but the following rules should be observed:

1. Less feeding should be done during dull weather and in the winter when the light intensity is low. Less feeding is required under cool conditions than in a heated greenhouse.
2. Sufficient liquid should be applied to soak right through the ball of soil.
3. Liquid manure should be given when the plant is in need of stimulation, but do not apply it too often. Only repeat the dose when it is evident that the plant requires further nutrients.

Commercial growers usually liquid feed regularly at every watering when plants are established. Feeding should be anticipated rather than wait for deficiency in every case.

JOHN INNES FEED. This is obtainable both in dry and liquid form, and is recommended to those who use the John Innes compost. It is normally applied at the rate of ½–1 oz. per gallon of water (preferably soft) and used straight away. The dry feed has the same formula as the liquid but with the addition of a 'carrier'. The dose is ½–1 teaspoonful per 6 in. pot. In the case of the dry feed, growers will probably find it most convenient to buy the made-up feed rather than mix it themselves.

LIQUID MANURE. There are a number of liquid manures available, and I use two – one called Farmura and another called Marinure. They are both organic in origin. They can be bought ready for dilution and the instructions governing their use are on the label.

Farmura (General) contains 9% nitrogen, 6.6% phosphates and 4.1% potash and should be diluted in the ratio of 1 teaspoonful to 1 gallon of water. For young plants, use once a week after transplanting until the first truss has set, diluting it in the ratio of 1 teaspoonful to 1 gallon of water. After the first truss has set, use 2 teaspoonfuls per gallon and feed twice a week giving 2 pints to each plant. When four trusses are showing fruits you can increase the dose to 4 teaspoonfuls per gallon, feeding two or three times a week and giving 2 pints per plant, increasing to 6 pints when the fruit start to ripen.

Marinure normally contains 5.75% nitrogen, 6.5% phosphates and 7.4% potash.

CARBON DIOXIDE. After about fifty years of experimentation in Great Britain, the Netherlands and the USA, a new technique – which may be called enrichment of the air with carbon dioxide – has become accepted as a part of greenhouse management. CO_2 (carbon dioxide) enrichment is capable of producing crop increases of up to 20–30% if gas of sufficient purity is used under first-class conditions.

The micro-organisms in any ordinary soils will produce carbon dioxide at a rate of only 2 to 3 lb. per acre per hour, but in cases where animal manure or straw bales are used for, say, cucumber beds, CO_2 may be produced at ten times this rate. This is enough to satisfy the demand by the crops at high light intensities. Thus some gardeners have used the CO_2 enrichment plan quite unintentionally for a large number of years! However, the rate of production of CO_2 from such organic materials varies, and of course does not allow of day to day control. Gardeners have therefore had to turn to other sources. It was once the custom of head gardeners to urinate down the paths of greenhouses at night-time. They knew this was of benefit to the plants but did not realise it was because of the extra CO_2 given off by the urine.

Propane and some good grades of paraffin are fuels which, when burned, produce CO_2 of sufficient purity for use in greenhouses. Propane is nearly twice as costly as paraffin, but has a lower sulphur content. Some heat is produced by this process. Almost pure CO_2 may be obtained as a liquid in cylinders or as solid 'dry ice'. The cost is two or three times as great, depending on the quality bought, as when propane is used, but the pure gas is more convenient and there is little risk of crop damage.

Under ordinary conditions, the proportion of added carbon dioxide taken up by the plants is smaller than the amount lost by ventilation. Distribution must therefore be designed to minimise this wastage. When the ventilators are closed and the heating system operates, convection currents distribute the CO_2 throughout the volume of air. It therefore does not matter whether the sources are few or many. The maximum benefit from CO_2 is obtained if it is accompanied by higher air temperatures than usual. For this purpose the ventilator should be left closed for a number of days in the early spring.

When in the spring and summer it is necessary to leave the ventilators open for much of the day, most gardeners discontinue their CO_2 enrichment. It is, however, under these conditions of higher sunshine that a rise in the concentration of CO_2 results in a great increase in photosynthesis. In the case of a large greenhouse, it is better to distribute numbers of small open containers of solid gas over the floor of the greenhouse.

Assimilation of carbon dioxide during photosynthesis is the process that enables plants to grow. Increasing the CO_2 concentration in the air around the leaves up to 1,000 v.p.m. greatly increases the rate of assimilation – but there is no point in exceeding this level. If the light intensity or the temperature is too low, raising the CO_2 concentration above the normal 300 v.p.m. is of no avail. The enrichment with carbon dioxide enables greenhouse plants to make more efficient use of light, and this naturally brings about changes in plant management.

The construction of the greenhouse itself may have to be slightly altered. The leaks between the glass overlaps in modern greenhouses are not accidental, for when the ventilators are closed they admit carbon dioxide from outside to replace that taken up by the plants. When CO_2 is given artificially, however, these gaps are no longer necessary, and the more tightly the house is sealed, the smaller will be the wastage and the simpler the problem. The need to keep the ventilators closed when CO_2 enrichment is in use will no doubt keep gardeners alert to new developments.

Carbon dioxide enrichment can be brought about in a totally safe and acceptable way merely by burning Propagas in the greenhouse. Tomato, cucumber and lettuce yields can be increased by 38% – some even by 60%. Chrysanthemums and carnations will increase their production by 38% and will show proportionate gains in head size. A small generator, specially designed for the small greenhouse, can supply sulphur-free CO_2. Further details can be obtained from Shell-Mex House, Strand, London.

PART THREE

Plants for the Greenhouse

CHAPTER 9

Growing Annuals in Pots

ONE of the simplest and cheapest ways of having a glorious show of colour in the spring, summer and autumn is to grow a series of different annuals in pots. In the Good Gardeners Association's Garden at Arkley, Herts., there is a double-glazed greenhouse which each year is filled with annuals in pots and the display is greatly admired by the thousands of visitors.

It is possible either to sow the annual seeds in boxes of Levington or Alexpeat compost, and then to plant out the little seedlings into their individual pots later, or to sow two or three seeds in the centre of the pot in which you wish the plant to grow and, if all germinate, thin down to one plant per pot later on. Either way, the plan is to grow the specimen on slowly, without any forcing at all. The taller plants need a little support – a split bamboo suits the purpose well, especially if dyed green with Rentokil preserving fluid. The tying is best done with a material called green twist or twill.

Annuals are easy and economical to grow, requiring little or no fire heat. A selection of the best ones is given on the following pages.

Antirrhinum *Antirrhinum majus* (Snapdragon)

The many antirrhinums which are suitable for pot culture come within the dwarf compact species, the best varieties being Golden Queen, Scarlet Queen, White Queen and Amber Gem, and the group is generally known as Tom Thumb. All these varieties branch freely and produce plants about 9 in. high. Larger species are unsuitable for pot work.

'Strain' is of more importance than variety in antirrhinums, and once the gardener has obtained a really good stock, it is usually advisable for him to carry out propagation by vegetative means.

Cultivation. Propagation is best done by means of cuttings taken from stock plants during July and inserted into a sandy compost in a cold frame. The cuttings may be either nodal or internodal and should be placed about 4 in. apart. As soon as the cuttings are well rooted and growing freely (September), they should be stopped (i.e. the growing points should be pinched out), and should then be potted directly into 6 in. pots, using Levington or Alexpeat compost No. 2.

The pots should be placed on the staging of a light airy house, allowing sufficient room for the leaves not to touch. A temperature of 50°F should be maintained throughout the growing period, ventilation being given on all possible occasions. The plants should be in bloom during late February or early March.

Feeding. Pick out the dry plants each day and water them. Apply a good liquid fertiliser just before flowering. Other feeds may be given if necessary before this.

Balsam *Impatiens balsamanica*

Cultivation. Sow the seeds in early March $\frac{1}{8}$ in. deep in Alexpeat seed compost or Levington compost in a seed box or in a 6 in. pot. Put the boxes or pots on the staging of the greenhouse at a temperature of 65–70°F. When the seedlings are 1 in. high, pot them up into 3 in. pots filled with the No-Soil compost No. 1 or Levington compost. Pot fairly firmly only, and after a watering stand the pots on the staging of the greenhouse and grow the plants on slowly at about 60°F. Be sure to give the plants as much light as possible – never apply any shading to the glass at this time.

Pot up from the 3 in. pots to 6 in. ones using the Alexpeat compost No. 2, or Levington compost, this time making the compost a little firmer. The plants can now grow on the staging at a temperature of from 50–55°F. Balsams need regular watering.

Feeding. Apply diluted Marinure until the plants are in full flower.

Varieties.

F.1. Hybrid Imperial Mixture – remarkable hybrid vigour, high degree of better quality flowers, and good germination. Colours include orange, carmine, rose and purple
Carmine Baby – bright orange-scarlet
Salmon Princess – bushy plants, salmon-rose
Scarlet Baby – bright scarlet, flat habit of growth

Clarkia *Clarkia elegans*

This can be grown at the same time as the calendulas (see p. 86).

Cultivation. Sow the seed very thinly during September in a seed tray containing Levington seed compost. Stand the box on the staging of the greenhouse in a temperature of 50°F. As soon as the seedlings are through, the box should be watered with Cheshunt compound as a

Aphelandra sinclairiana.

Above left, a fine, hanging basket of *Campanula isophylla*; *right*, *Chlorophytum elatum* 'Variegatum' under the staging. *Below*, *Abutilon megopotamicum* 'Cynthia Pike'.

safeguard against damping off, for once this disease gets a hold it will soon sweep right through the plants. As soon as the plants have made about four leaves, they should be potted off singly into 3 in. pots, using Alexpeat or Levington compost. The pots are set out upon the staging 2 in. apart. Care must be taken not to bruise the stems, for when this happens root rot is set up. Watering must be kept to a minimum. During January the plants are potted on into 6 in. pots, this time using Alexpeat or Levington compost No. 2.

When the plants are about 5 in. high, any that have not 'broken' naturally should be stopped, i.e. the growing point should be pinched out. Cold draughts should be avoided; as the weather becomes colder, ventilation should only be given on the leeward side of the greenhouse, and a temperature of 45–50°F should be maintained throughout the life of the plants. One thin stake is needed per plant, and all the side shoots are then looped in with raffia or green twine as they develop.

Feeding. During the last three weeks a feed of liquid manure should be given every five days to ensure that the foliage is of a good green colour.

Varieties. The doubles are considered to be best:

Salmon Queen
Scarlet Queen
Purple Prince – for some reason, this one is not particularly popular

Celosia *Celosia cristata* (Cockscomb)

This is an attractive plant with red or yellow 'cockscombs' which look like coloured pampas grasses.

Cultivation. For July flowering, seed should be sown during March in trays of Levington or Alexpeat seed compost, and placed in the greenhouse at a temperature of 75°F. As soon as the seedlings are large enough they should be pricked out into further seed trays, 54 per tray, using a No-Soil compost No. 1. When they are well rooted, pot them off into 3 in. pots, again using a potting compost No. 1, and as soon as the cockscombs are developing they should be potted on into 6 in. pots – this time using a compost No. 3, and be placed 8 in. apart on the greenhouse staging. Care must be taken to see that the plants do not become pot-bound or they will flower prematurely. A humid atmosphere and a fairly high temperature (65–70°F) are necessary.

The plants should be watered carefully until well-rooted, after which water freely. Slight shade may be necessary to prevent leaf scorch in sunny weather.

Feeding. Give Marinure at two-week intervals during the growing period.

Species and varieties

Celosia argentea cristata – various colours
C. pyramidalis var. Thompsonia
Jewel Box – particularly beautiful

Cuphea *Cuphea platycentra* (Cigar Plant)

Really a perennial, but when sown early it flowers the first year. The plants are freely branching and bear numerous little scarlet flowers with white tips which are said to resemble miniature lighted cigars. It is a plant which is particularly suited to pot culture.

Cultivation. The seeds are best sown in early February in Alexpeat or Levington compost in a seed box or 6 in. pot. Cover them lightly with a little silver sand, press down and water by immersion. Put the pots on the staging of the greenhouse at a temperature of 65°F.

When the seedlings are through and about half an inch high, pot them directly into 6 in. pots filled with Alexpeat or Levington compost No. 2. Crock well if clay pots are used, but this is not necessary with plastic polypots.

The aim should be to have the plants in flower from about the middle of May until October, and during this period the plants will want watering moderately; the greenhouse temperature at this time can be as low as 50°F. The plants should grow bushily of their own accord reaching about 1 ft. high, and should not need staking.

Carnation *Dianthus caryophyllus*

Cultivation. Sow the seed thinly in March in a seed box filled to within half an inch of the top when the soil is pressed down level. Use Levington or Alexpeat compost, and after sowing cover the seed shallowly with a sifting of coarse sand. Press this down and water the box by the immersion method, dipping very slowly. Put the seed box on the staging of the greenhouse at a temperature of 55°F. Cover with a sheet of glass and a sheet of brown paper, removing them each day to wipe the glass. When the seedlings are through remove glass and paper altogether.

When the seedlings are 1 in. high pot them up with the Alexpeat Compost No. 1, and after three weeks pot on again into 6 in. pots, three seedlings per pot. If preferred you can grow on the plants in 3 in. pots. Disbud the blooms, leaving the top flower-bud in position and removing the side-buds.

Varieties

Dwarf Double Pygmy – 9 in. high, fully double, dwarf, bushy compact plants. Available in different colours
Grenadin – 12 in. high, early flowering. Available in different colours
Enfant de Nice – 20 in. very attracitve, different colours
Margarita – 15 in. free-flowering, excellent flowers in bloom within five months of sowing seed

Exacum *Exacum affine*

This plant produces bluish-lilac fragrant flowers on attractive stems 6 in. high, covered with glossy green leaves. It is possible to have pot plants available and in flower from May to October.

Cultivation. The seeds should be sown about the beginning of March, in Alexpeat or Levington compost, either in a seed box or in a 6 in. pot. Cover the seeds with a little fine compost sifted through an ⅛ in. sieve. Press this down lightly and water the box or pot by immersing it very gradually in tepid water until the moisture is seen to rise up to the surface. Put the pot or box on the staging of the greenhouse at a temperature of 75°F.

When the seedlings are about ½ in. high, transplant by placing each one very carefully into the centre of a 3 in. pot and firming the compost with two fingers of one hand. This time use the Alexpeat or the Levington compost No. 1. The plants should now be grown on the staging of the greenhouse at a temperature of 60°F. If desired they may be kept in these pots all the time, but I prefer to pot them on once more into 6 in. pots a month or six weeks later, this time using Levington or Alexpeat compost No. 2. In the 6 in. pots they make beautiful masses of scented lilac-blue flowers.

Godetia *Godetia amoena*

The godetia is an attractive flower similar in shape and size to the Evening Primrose. It is obtainable in various shades including white, pink, red, and red and white. The bushy species is the best for pot culture:

G. WHITNEYI AZALEAFLORA

Cultivation. Seeds should be sown during March or September in seed trays filled with Alexpeat or Levington compost to within ½ in. of the top.

When the plants are large enough they should be potted off into

3 in. pots, say three to a pot, using the Alexpeat or Levington potting compost No. 1. Maintain a greenhouse temperature of from 45–50°F at this time. As soon as a good root system has been established, the plants should be potted on into 6 in. pots using potting compost No. 2. The pots should be put 6 in. apart on the staging, in continuing cool airy conditions. Water each day, picking out and watering those pots which have actually dried out. A short stake may be given to any plants which appear to need support.

Feeding. Feed with Farmura at three-week intervals when the plants are growing well. Do not overdo feeding as this will result in coarse bushy growth and no flowers.

Varieties

Crimson Glow	Lavender Queen
Salmon Princess	Scarlet Emblem

Heliotrope *Heliotropium arborescens*

This is the most valuable fragrant pot plant; really a little shrub but first-class when grown as an annual. It is commonly called Cherry Pie, and bears mauvish-blue and white flowers all through the summer and well on into the winter. It does best in a sunny position.

Cultivation. Sow the seeds late in February or early March, $\frac{1}{16}$ in. deep in Alexpeat or Levington compost in a well-drained pot or box. Cuttings may also be taken in the Autumn or Spring.

Having sown the seeds very thinly, the plants may be allowed to grow until they are 1 in. high, when they should be potted up directly into 6 in. pots filled with Alexpeat or Levington compost No. 2. Pot fairly firmly, pressing the compost down with two fingers of one hand. Put the plants on the staging of the greenhouse at a temperature of 60°F and water freely from March to October. If you want dwarf bushy plants, pinch out the points of the main and lateral shoots when they are 3 in. long, and do another pinching back when the secondary shoots are 6 in. long.

Feeding. When the plants start to come into flower, feed with diluted Marinure once a week.

Varieties

Royal Marine – deep violet
Regale – mauve

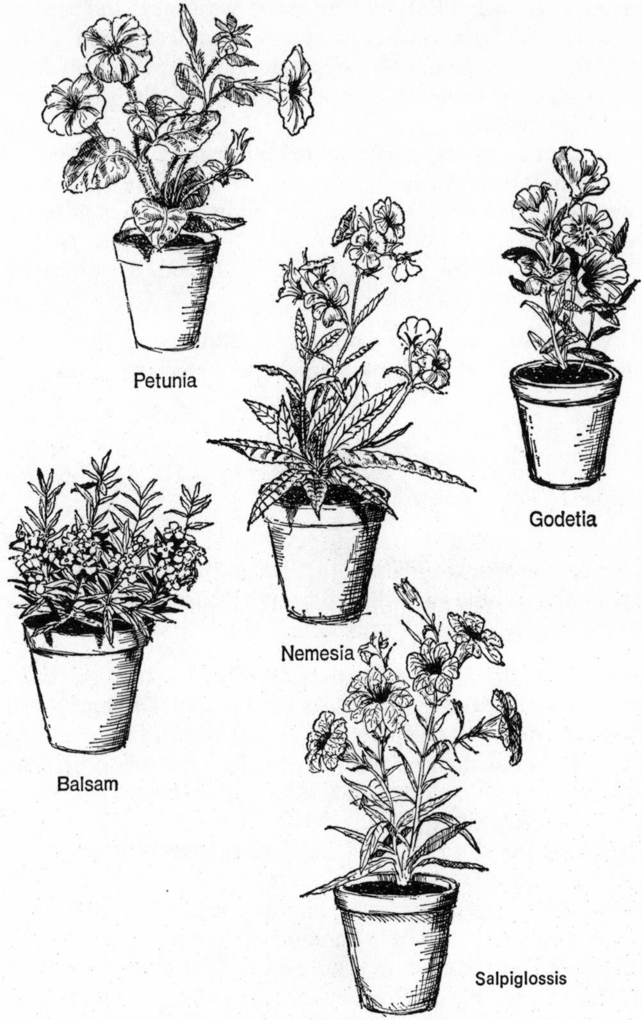

FIG. 21 Five annuals in pots

Jacobaea *Senecio elegans*

This plant is extremely easy to manage and is very useful as a 'stop-gap'. Some people consider it looks rather like groundsel, but its pink, white or mauve flowers in April and bright green foliage do, I think, present a very pretty sight.

Cultivation. Sow seeds thinly during early September in standard seed trays filled with Alexpeat seed compost, at a temperature of 50°F. When large enough to handle the seedlings should be potted off into 3 in. pots of Alexpeat or Levington compost No. 1, and placed 2 in. apart on the greenhouse staging.

In February the plants should be ready for the final move into 6 in. pots of Levington or Alexpeat compost No. 2. The pots should be placed on the staging of the greenhouse, allowing sufficient room so as to prevent the leaves from touching. Any 'swingers' (i.e. those plants which have weak floppy stems) should be staked, as otherwise they will never make good plants.

Feeding. When the plants have become established in their new pots, they should be given fortnightly waterings of weak soot water.

Varieties

Tall Double Mixed – various colours

Marigold *Calendula officinalis*

This is one of the most popular annuals for pot work, flowering from mid-March onwards.

Cultivation. Sow the seed early in September in seed trays filled to within half an inch of the top with Levington or Alexpeat seed compost, and cover to a depth of a quarter of an inch with sifted compost. A sheet of glass should then be placed over the boxes, and brown paper put over the glass to exclude light. The paper should be removed as soon as germination takes place.

The boxes should be placed in the greenhouse at a temperature of 45–50°F.

As soon as the seedlings show their first rough leaf, they should be potted off singly into 3 in. pots containing Levington or the Alexpeat compost No. 1, no crocks being necessary. The pots should then be set out upon the staging 2 in. apart, and be given plenty of ventilation when the weather is favourable. The plants will be happy in these pots until the end of January, and only the minimum of watering should be given during the very short days. The plants should then be potted on into 6 in. pots using Alexpeat or Levington compost. Too rich a compost will encourage a lot of soft leafy growth and few flowers. Stop each plant when it is 4 in. high, as otherwise there will be only a solitary bloom arising from the terminal bud. All weak shoots should be removed if they appear, or else the main shoots will suffer.

Varieties

Geisha Girl
Camp Fire Improved
Golden King – various shades of orange
Indian Maid
Monarch Orange

Mignonette *Reseda odorata*

Mignonette is an extremely sweet-scented plant and very popular for indoor decoration. It is rather liable to stem canker which is sometimes difficult to control, but which can be overcome by careful cultural operations.

Cultivation. Sow seeds in early September in 3 in. pots (two or three seeds to each pot) filled with an Alexpeat or Levington seed compost. The pots should be put upon the staging of an airy greenhouse, close together, at a temperature of 50°F. As soon as the seeds have germinated they should be reduced to one plant per pot, and from this time onwards a little hydrated lime suspension in water should be given every ten days. This helps to prevent the stem canker attacking the plants.

When the plants are 3 in. high they should be stopped, and then be given a little more space on the staging so that the leaves do not touch and to allow free circulation of air around the plants. Two side-shoots should be retained, all others being removed. When these shoots are about 5 in. long they should be stopped again. This time three of the sub-laterals should be retained, thus producing a plant with six shoots. The plants should be ready for potting on into their final pots early in the new year, using the Alexpeat potting compost No. 2 or the Levington compost. The pots should be stood on the staging of the house at a temperature of 50°F, ventilation being given whenever possible. Care should be taken when watering, especially early in the season.

This plant makes a better specimen when grown in 3 in. pots. If, however, 6 in. pots are to be used, it is advisable to increase the number of shoots by allowing further laterals to develop, and probably increase plants to 3 per pot.

Varieties

Orange Queen
Red Monarch – deep red

Nemesia *Nemesia strumosa grandiflora*

This plant is very pretty during March or April, and will give a good mass of colour when grown as a pot plant.

Cultivation. The seed should be sown thinly in September in standard seed trays filled with Alexpeat or Levington seed compost. The boxes should be placed either in a cool airy house or in a cold frame, and a light watering should be given.

As soon as the plants are big enough to handle, they should be potted off in clumps of four to six plants into 3 in. pots, where they should remain until they have formed a good root system. The house temperature should be kept at 50–55°F, allowing plenty of ventilation on all favourable occasions. Potting into the final 6 in. pots should be carried out early in the New Year using Alexpeat or Levington compost No. 2. Great care must be taken when potting as the plants will have formed a good root system, and frequent watering is necessary.

A short stake may be necessary if the plants become at all floppy as they grow.

Varieties

Blue Gem – blue
Fire King – red
Orange King – orange
Monarch Carnival – mixed colours

Petunia *Petunia grandiflora*

For pot work it is convenient to grow the dwarf varieties, which can be divided into three groups: dwarf bedding 12 inches, bedding 15 inches and tall 18 inches. The flowers are trumpet-like.

Cultivation. Sow the seeds in February in Alexpeat or Levington seed compost in seed boxes or pots at a temperature of 65°F. Sift only a very little compost over the seeds to cover them and having pressed this down, water carefully by the immersion method. Put the pot or box on the staging of the greenhouse. When the seedlings are 1 in. high, pot them up into 3 in. pots using a potting compost No. 1, allow the plants to grow undisturbed for six weeks, and then pot them up into a potting compost No. 2, this time into 6 in. pots. Pinch back the growing points of the plants so as to make them bushy, and a fortnight later start feeding with Marinure, applying a weak solution once every ten days until the plants are in flower. Aim to have the temperature at 55–60°F.

Varieties

DWARF BEDDING
Celestial – rich deep rose
Fire Chief – brilliant scarlet-red
Lavender Queen – silver-lilac
Violet Queen – beautiful violet
Rosy Morn – brilliant rose, with a large white throat

BEDDING
Blue Bee – clear violet-blue, a most outstanding colour
Crimson King – crimson-magenta
Howard's Star – reddish purple, white star
Rose King – rich rose, white throat

TALL
Superbissima mixed – giant flowers in a magnificent range of rich colours, all with veined throats
Dwarf Giants of California Mixed – extremely early, dwarf and large-flowered

Stocks *Matthiola* (East Lothian)

Pink and scarlet East Lothian stocks are very popular. They grow 9 in. high in 6 in. pots, and each plant will produce about three to four spikes of flowers.

Cultivation. In August, sow the seeds (two per pot) in well crocked 3 in. pots of Levington or Alexpeat seed compost, and place the pots in a cold frame until the end of September when they should be transferred to a light airy house. Here they should be stood about 4 in. apart each way and kept fairly dry throughout the winter. In February the seedlings should be potted on into 6 in. pots using Alexpeat or Levington compost No. 2 and set out on the staging of the greenhouse, allowing sufficient room to prevent the leaves touching. The temperature should be maintained at about 45–50°F, giving plenty of air on all possible occasions.

It is impossible to tell which plants will be double or single during the seedling stage, but those with a white 'farina' – a mealy dusty substance – on the leaves are far more likely to be doubles than the green-leafed ones.

Varieties

Apple Blossom – pink
Carmine Rose – carmine
Lilac Rose – rose
Rose Queen – deep rose
Dark Blue – blue

Salpiglossis *Salpiglossis sinuata*

This is a half-hardy annual sometimes called the Scalloped Tubed Tongue, because of its trumpet or tube-shaped flowers.

Cultivation. Because the plants are particularly valuable for making a brilliant show in February, March and April, the seeds should be sown in July and August. It is possible to do this sowing out-of-doors in a sunny position in drills 6 in. apart, and ½ in. deep. When the seedlings have formed three leaves, they should be forked up carefully, ready for potting up. It is advisable to water the beds thoroughly two days in advance so that when the seedlings are forked up, the leaves are really firm and turgid. Pot up the seedlings fairly firmly in Alexpeat or Levington compost No. 2 in 3 in. pots, which will need crocks unless of plastic, give a light watering, and place on a shelf close to the glass of the greenhouse, at a temperature of 55°F. When the plants are really well rooted, pot them on into 6 in. pots using the Alexpeat compost No. 3 or Levington compost, and place on the staging of the house. The potting on this time should be a little firmer.

When the main shoots are 6 in. high, pinch out the growing points to encourage bushy growth. The plants should remain in a cool greenhouse until they flower, when they may be brought into the home if desired and placed on a window-sill near the light.

Varieties

Monarch Mixed F.2 Hybrids – large flowers in a brilliant range of colours. Dwarf compact plants

Grandiflora Superbissima Mixed – compact heads of large open flowers of the richest colours

Schizanthus *Schizanthus hybridus grandiflorus* (The Butterfly Flower)

A very decorative plant producing extremely pretty flowers which from a distance look like butterflies. They may be mottled pink and white and striped in chocolate and brown. This plant needs quite a high temperature in its seedling stage, but this may be reduced during the actual growing period.

Cultivation. The seed should be sown thinly during February and March in standard seed trays filled with the Alexpeat or Levington seed compost. The boxes should be watered and then placed on the staging of the greenhouse at a temperature of 60–65°F. As soon as the seedlings are large enough to handle, they should be potted off singly into 3 in

FIG. 22 Pot of Schizanthus

pots, using the Alexpeat or Levington potting compost No. 1, and set out on the staging 2 in. apart, in a temperature of 50°F. Great care should be taken with watering – in fact in the early stages the pots should be kept on the dry side.

As soon as the roots have penetrated the ball of soil the plants should be potted on into 6 in. pots, using Alexpeat or Levington compost No. 2. Again care should be taken with watering until the plants are really well-rooted.

Schizanthus should be stopped once or twice during growth in order to ensure a good bushy plant. One central stake is necessary, and, tied with green cotton twill or twine, is quite sufficient to support the plant. If the weather is extremely sunny, light shading is necessary to prevent the plants from becoming hard and yellow.

An autumn sowing (during September) will produce plants in flower during March and April.

Varieties

Crimson Cardinal – crimson
Dwarf Bouquet Mixed – mixed colours
Doctor Badger's Hybrids – mixed colours
Cherry Shades – cherry-red

Nasturtium *Tropaeolum majus*

The Tom Thumb nasturtiums are quite useful to fill an odd corner in the greenhouse.

Cultivation. Treatment should be similar to that for godetias, the only difference being that seeds are sown directly into 3 in. pots, three per pot. It is inadvisable to feed nasturtiums unless the leaves become yellow. The pots should be put on the staging 6 in. apart. Great care is needed in watering: pick out the dry pots each day and water these only. A short stick may be necessary in some cases to prevent the plants from flopping over.

Varieties

Empress of India – red and yellow

Thunbergia *Thunbergia alata* (Black Eyed Susan)

This plant comes from tropical Africa, and produces deep orange flowers on 4–6 ft. tall plants the whole of the summer. Can be used as a climbing or hanging plant.

Cultivation. Sow seeds thinly in late February or early March in Alexpeat or Levington seed compost in a well-drained pot or box, at a temperature of 65°F. Give a little water and put the pot or box on the staging of the greenhouse where there is plenty of light. When the seedlings have produced three good leaves, pot them up fairly firmly into 3 in. pots using a potting compost No. 2 and grow on quietly on the staging of the greenhouse at a temperature of 60°F. Six weeks later, pot on once more into 6 in. pots, using the potting compost No. 2.

When the shoots get too long, place the pots at the edge of the staging and allow the growths to hang down. It has large deep orange flowers, with decorative buds. The orange lanterns give a bright display for many months. Some people prefer to pot the plants up into wire or plastic hanging baskets and suspend these from the roof. Whichever method is used, the plants should be watered freely.

Varieties

Orange Lanterns – orange

Zinnia *Zinnia elegans* (Youth-and-Old-Age)

The zinnia is a half-hardy annual, well-suited to the cool greenhouse, with attractive large daisy-like flowers with flat centres.

It is possible to grow these plants either as single-stemmed unstopped plants with one huge flower, or as a bush (by stopping the plant when it is about 6 in. high) with a number of smaller flowers produced from the lateral buds which grow out from the leaf axils.

Cultivation. Seed should be sown during January or February in the greenhouse, in an Alexpeat or Levington seed compost, maintaining a temperature of 50–55°F. Seed trays should be filled with the compost to within half an inch of the top when pressed down evenly with a wooden presser. Sow one seed every 2 in. on the soil surface, and press each one in a quarter of an inch with the point of a pencil.

When the seedlings are large enough to handle, pot them off into 3 in. pots, using the Alex or Levington compost No. 1. As soon as the plants have rooted well and have penetrated the ball of soil, they should be potted on once more into 6 in. pots, this time using the Alexpeat potting compost No. 2, or the Levington compost. The pots should then be put on the staging in the greenhouse at a temperature of 50°F, allowing 8 in. between the plants. A temperature of 50–55°F should be maintained throughout the growing life of the plants, taking care to give plenty of ventilation.

If any plants are weak they should be tied loosely to a small stake with twill or fillis.

Feeding. Marinure should be applied just as the buds begin to swell in order to ensure good healthy foliage, and this feed can then be given regularly, say every week or ten days, until the plants are in full flower.

Varieties. There are several good strains, including the Giant Dahlia-flowered, the *robusta grandiflora plenissima*, and the *pumila flore-pleno*. All of these are obtainable in many lovely shades, including brick-red, rich yellow, rose, apricot, bright scarlet, deep purple, cream and white. Newer strains and varieties include Pink Buttons, Red Buttons, Gaillardia Flowered, Canary Bird, Purple Prince, Scarlet Flame and Golden Dawn.

CHAPTER 10

Flowering Perennials in Pots

SOME people prefer foliage plants to flowering plants because on the whole they are easier to look after. Furthermore, many of them can be taken into the home to grow in a shady corner and form a permanent display. Flowering house-plants need more careful management and generally flower only during certain times of the year, but they do provide a varied and attractive range of colours both in the greenhouse and indoors.

Some writers have tried to divide flowering pot plants into sections such as spring, summer, autumn and winter, but to me this seems rather unwise, because there is a tremendous amount of overlap and by potting up later or by applying less heat, plants may bloom a month or two after 'the book' says they should do.

As to general management, there is very little difference between the foliage plant and the flowering plant except perhaps that when the plant starts to flower it is usually advisable to stop syringing it over. Flowering plants like the compost to be kept just moist, especially in the spring and summer, and a few hours' drought may cause a great deal of harm. Soddenness and over-dry air are also disliked, and it is quite a good idea on hot days to water the greenhouse path thoroughly as well as to give a good watering underneath the staging. I often do this early in the morning and again in the later part of the afternoon in June, July and August. Dry heat and the over-dry leaf encourage red spiders and thrips, and the flower-buds on many plants like gloxinias and begonias may fall off. Syringing the undersides of the leaves carefully once a day in the summer is most helpful, and in really hot weather it pays to syringe early in the morning and late in the afternoon as well.

Though one must never keep out all the sunlight, it does help to shade from sunshine and many beginners fail to recognise the difference between the two. Strong sunshine may scorch and burn, and if you provide an outside blind of hessian or slatted cedar strips it helps greatly in keeping flowering plants at their best. Towards autumn, as the days get shorter, the shading may gradually be removed until, perhaps in October, the plants are growing in full light. It is often necessary to start shading again in May in the south, and probably early June in the north.

It must be remembered that in this chapter we are concerned with what we may call potted perennials. There are of course other flowering plants which can be grown in the greenhouse, and these are described in other chapters in this section, e.g. annuals in Chapter 1, the chrysanthemums and carnations in Chapters 4 and 5, orchids in Chapter 6, and bulbs and corms in Chapter 7.

It is obviously impossible in a book of this size to mention every possible species or variety of plant which could be grown in the greenhouse. What I have done here is to include (a) those that are easiest to grow, and (b) those that can be obtained without much difficulty from nurserymen in this country. I have grown all of them at some time or another – either in my own greenhouses or in the greenhouses I managed, such as those of the Swanley Horticultural College and the Cheshire College of Agriculture – and often I have been able to include special hints on the management of some plants as the result of personal experience.

The following, then, are the plants I would recommend. It will be impossible for most readers to attempt to grow all the plants I have described, let alone the various species and varieties. The descriptions will, however, enable readers to make their own choice – and in cases where a plant is to be given as a present, details of the correct management may also be supplied without difficulty.

ABUTILON

A plant with lovely foliage and drooping bell-shaped flowers from April to July. I have grown mine out in the garden in mid-summer.

Cultivation. Sow seed in February at 65°F for plants flowering in early July. For the first year, grow in 6 in. pots of Levington or Alexpeat compost No. 3, and re-pot into 8 in. pots in the second year, in March. Water freely when it is hot and, when the plants are growing well in summer, feed weekly with Marinure. Water moderately in winter. The greenhouse temperature should be 50°F. Prune quite hard in February. Propagation is also possible by 2 in. long cuttings, struck in the propagating frame at 65°F.

Species

A. savitzii – green and silver leaves
A. thompsonii – green and gold leaves
A. darwinii var. Golden Fleece – yellow flowers
A. d. Jubilee – pink
A. d. Firefly – crimson

AGATHAEA – see Felicia

ALPINIA (Indian Shell Flower)

A very showy plant flowering in spring and summer. The rhizomes smell of ginger.

Cultivation. Propagate by division in March or April, potting into Levington or Alexpeat compost No. 3. Re-pot into 6 in. pots in March. Grow on the greenhouse staging at a temperature of 70°F from March to September, and 55°F for the rest of the year. Water freely in summer, giving Marinure once a week (without this, the flowers are poor), and moderately in winter.

Species

A. rafflesiana – small flowers which are yellow with red tips
A. r. sanderae – shiny green leaves with broad white bands
A. mutica – flowers white with spotted orange lip

AMASONIA

A lovely evergreen flowering in September.

Cultivation. Pot in Levington or Alexpeat compost No. 2, re-potting in March. Grow on the greenhouse staging at 70°F from March to September, and 55–60°F for the rest of the year. Grow in full light except in mid-summer, when some shading is necessary. Water well in spring and summer.

Species

A. calycina – pale sulphur-yellow with red-purple hairs and red bracts
A. erecta – white and pink flowers, scarlet bracts

ANTHURIUM

Not an easy plant to grow but included because of its great beauty. The foliage is pretty and the flowers brightly coloured and flat, with a spathe or 'tongue' sticking out of the centre.

Cultivation. Sow seed in the greenhouse or propagating frame at 80°F in well crocked pots (good drainage is essential) in Levington or Alexpeat compost No. 3. Re-pot in March. Grow in a semi-shady spot at 55°F. Moss may be grown around the base of the plant to conserve moisture.

Above left, 'Bridal Gown', a chrysanthemum of the reflex form; *right, Senecio cruentis. Below,* perpetual-flowering carnations.

Above, the white fuchsia 'Evensong'. *Below, Clerodendron thomsonae.*

Species

A. andreanum, Flamingo Plant – reddish-orange or scarlet. Has many hybrids, such as the salmon-pink Dr Lawrence, listed as varieties

A. scherzerianum – brilliant scarlet. Many hybrids, such as Cardinal, a crimson-scarlet

APHELANDRA

Blooming in the autumn.

Cultivation. Propagate in April by cuttings of half-ripened wood in Levington or Alexpeat compost No. 3, giving bottom heat. Re-pot in March. Temperature: 70°F from March to September, 60°F for the rest of the year. Water freely in summer, moderately in winter, keeping the atmosphere in the house moist. Feed with Marinure when flower spikes are first seen, and continue once a week until the flowers are fully out. Prune back the shoots in February if desired.

Species. There are a great many; these are the easiest to grow:

A. squarrosa louisae – dark leaves with white veins, 'square' yellow flowers

A. leopoldii – smaller than *squarrosa*, thinner and shorter leaves which are light green with a silver zone

ARDISIA (Spear Flower)

These plants always look their best when just 2 ft. tall; after this they tend to lose the bottom leaves. After flowering in June they may become covered with scarlet berries.

Cultivation. Propagate in April, May or June by striking 2 in. long cuttings of half-ripe wood in Levington or Alexpeat compost No. 2 in the propagating frame, or by seeds sown in March or April in the propagating frame at 75°F. Give plenty of water during summer, but only a little in winter.

Species

A. crispa – reddish-violet scented flowers

A. macrocarpa – flesh-coloured flowers followed by vermilion berries.

ASCLEPIAS (Milkweed)

A twining plant flowering in July, August and September and related to Ceropegia, a plant sometimes grown for hanging baskets.

Cultivation. Propagate by division of the roots in October or April, or by seed in spring. Pot in Levington or Alexpeat compost No. 2 and re-pot, if necessary, in April. Water well from spring onwards and syringe over regularly. When the stems are 6 in. long, pinch off the growing points to encourage bushy growth. A month later start feeding with Marinure and continue weekly until September. Cut back the plants in late autumn each year, and keep on the dry side in winter.

Species

A. curassavica – reddish-purple flowers
A. cl. alba – white

BEGONIA

The tuberous-rooted begonias are grown for flowers; for foliage species see p. 185. There are also some fibrous rooted begonias which can be raised from seed sowing.

Cultivation. Start off the tubers in March in damp sedge peat at a temperature of 60°F, making sure the tubers are the right way up. After the plants have started to grow, pot up into 6 in. pots of Levington or Alexpeat compost No. 2. Re-potting is not normally necessary. Stake the plants as the stems are weak, give plenty of ventilation, a temperature of 60°F, and do not over-shade. Syringe daily in summer until the plants are in full flower and feed weekly with Marinure when they are growing well. After flowering, gradually withhold water until the leaves turn yellow and shrivel, then store the corms in dry silver sand at 40°F during winter.

Species and varieties. The following are varieties of *B. multiflora*:

Flamboyant – vivid scarlet
Jewel – orange-yellow
Mayor Max – double orange-scarlet
Mrs Helen Harms – soft clear yellow
Mrs Richard Galle – dark foliage, gold and crimson flowers

BELOPERONE

A Mexican plant flowering from July to September, but occasionally persuaded into flower in May.

Cultivation. Propagate in February or March by 2 in. long cuttings inserted in a peaty sandy compost in the propagating frame at 70°F, or by seed. Pot in Levington or Alexpeat compost No. 1 and re-pot in March. Temperature: 60–65°F. Water well from May to September and pinch out the growing points of the tall shoots to keep the plant bushy.

Species

B. guttata, Shrimp Plant – purple and white shrimp-like flowers
B. violacea – violet flowers

BILLBERGIA

Evergreen plants with showy flowers in spring.

Cultivation. Propagate in April by rooted side shoots at 85°F. Pot in Levington or Alexpeat compost No. 2 and re-pot in March. Prefers a fairly sunny spot in the greenhouse, and a temperature of 65°F. Water well in summer, but take great care to avoid soddenness. Feed weekly with Marinure in summer.

Species

B. nutans – yellowy-green flowers with red bracts
B. zebrina – yellowy-green flowers, spotted and white-banded leaves

BRUNFELSIA

A Brazilian evergreen flowering in summer.

Cultivation. Propagate in spring by cuttings of new growth or well-ripened wood in the propagator at 65°F. Pot in Levington or Alexpeat compost No. 2, and re-pot only rarely, as the plants flower best when pot-bound. Likes a semi-sunny position, a winter temperature of 50°F and a summer one of 70°F. Water freely in summer, syringing daily and feeding once a week with Marinure, and moderately in winter. In summer when shoots are 6 in. long, pinch out the growing points. After flowering, thin out the shoots.

Species

B. calycina – scented purple flowers
B. americana – yellow and white flowers
B. latifolia – white or purple, winter

CALCEOLARIA (Slipper-wort)

Plants up to 3 ft. tall with large yellowish flowers marked with spots and blotches. Herbaceous species flower from May to October, shrubby species in summer.

Cultivation. Propagate herbaceous kinds by seed in July and the shrubby kinds by cuttings in spring. Pot in Levington or Alexpeat compost No.

2, re-pot in March and subsequently as required. When re-potting, prune back the shrubby species. Grow both kinds in semi-shade, at 50°F, and nip off leading shoots when 8 in. high to encourage bushy growth. Never over-water, but do not allow the compost to dry out in winter. Established plants should be fed once a week. Discard the herbaceous kinds after flowering.

Species

SHRUBBY
C. fuchsiaefolia – yellow flowers, April
C. integrifolia – yellow to red-brown flowers
C. mexicana – pale yellow bubble-like flowers

CALLISTEMON (Bottle-brush Tree)
A lovely Australian evergreen with flowers like bottle-brushes in June.

Cultivation. Propagate in August by 3 in. long cuttings in sandy peat in the propagating frame at 60°F. Re-pot occasionally, in March, as the plants flower better when pot-bound. Grow in full sun, give plenty of ventilation and a temperature of 55°F. Water well in summer, a little in winter. In the years when not potting up, give a top dressing of fresh compost.

Species

C. citrinus – crimson flowers
C. salignus – yellow flowers

CAMPANULA (Bellflower)
Flowering in summer.

Cultivation. Propagate in August by seed in the cold frame, or in spring by cuttings. In October transplant seedlings into 3 in. pots of Levington or Alexpeat compost No. 2 and re-pot into 7 in. pots in March (*C. pyramidalis* grows to 5 ft. and needs an 8 in. pot). Grows easily in a coolish house, about 50°F. Water well in summer. Watch out for mites and red spiders, and spray underneath the leaves two or three times a week in summer. Also spray with liquid Derris.

Species

C. pyramidalis – white or blue; only suitable for large greenhouses
C. isophylla – blue or white, tumbling over the front of the pot

1. *Abutilon* × *hybridus*, 'Red Ashford', p. 95
2. *Acalypha wilkesiana*, p. 183
3. *Begonia rex* hybrid, p. 185
4. *Beloperone guttata* (green form), p. 98
5. *Brunfelsia calycina* var. *macrantha*, p. 99
6. *Calceolaria*, p. 99

CANNA (Indian Shot)
Broad leaves and brightly coloured flowers in summer.

Cultivation. Propagate in March by division of the roots. Pot in Levington or Alexpeat compost No. 3 and re-pot in March. Likes a sunny spot on the greenhouse staging, and a summer temperature of 70°F and a winter one of 50°F. It can be stood outside in mid-summer. When growing well, give plenty of water through spring and summer, but after the end of September until, say, the start of April, give little water. Feed weekly with liquid manure from June to September.

Species

C. edulis – bright red flowers, 6 ft. The tubers are edible (hence the name)
C. indica – yellow and red flowers, 4 ft.
C. warscewiczii – scarlet tinged with blue, 4 ft.
Named varieties (hybrids) are also offered by nurserymen

CELSIA (Cretan Mullein)
A relative of the verbascums, flowering in May and June.

Cultivation. Propagate by cuttings struck in the propagating frame at 60°F, and pot up into 3 in. pots of Levington or Alexpeat compost No. 1 any time from early March to the end of May. Re-pot into 6 in. pots in March. Grow in the sunny part of the greenhouse at 60°F in summer and 45°F in winter. They do not need a lot of water at any time, but very little should be given in winter. Feed twice a week with Marinure in summer. It may be necessary to stake *C. cretica*.

Species

C. arcturus – yellow flowers with purple centres, very beautiful
C. cretica – 4 ft. long spikes of yellow flowers

CHIRONIA
A very useful but seldom grown S. African plant, flowering in August when few other plants are at their best. Best treated as a biennial.

Cultivation. Propagate by seed in June or by 2 in. long cuttings in summer. Pot up the seedlings into 3 in. pots of Levington or Alexpeat compost No. 3 and re-pot again into 6 in. pots in October. Always pot firmly. Grow in the sunny part of the greenhouse at 60°F in summer and 50°F in winter. Feed with Marinure once a week until the flowers appear. Keep the plants fairly dry from early October until late February.

Species

C. linoides – 12 in. high; pink circular flowers 1 in. across
C. floribunda – pink flowers and smooth dark green leaves

CINERARIA – see Senecio

CONVALLARIA (Lily-of-the-Valley)

There are three main methods of growing lily-of-the-valley: (1) forced, flowering from January to May, (2) natural growth outside, flowering June and July and (3) what are termed 'retarded crowns', flowering September to December.

Cultivation. Lift crowns from open ground and pot up into 6 in. pots (eight per pot) of Levington or Alexpeat compost No. 2. Plunge pots in ashes and keep these moist at 65°F. When the plants are about 2 in. high, remove the pots from the ashes and place on the greenhouse staging at 55–60° F. Provide a tiny bamboo and loop of raffia for each pot, and keep in semi-darkness until the spikes are 5 in. high and then move into full light, but keeping the house lightly shaded. Syringe regularly to maintain a humid atmosphere, stopping as soon as the flower spikes appear and reducing the temperature to 55°F. Autumn and winter flowering plants can be obtained by buying retarded crowns.

Species and varieties

C. majalis – scented white bell-like flowers
C. Fortin's Giant – very large flowers

CUPHEA (Cigar Plant)

Easy to grow, flowering in June and July.

Cultivation. Propagate in March or April by striking 2 in. long cuttings in the propagating frame at 65°F. Pot in 6 in. pots of Levington or Alexpeat compost No. 1, and water well through summer, but only moderately in autumn and winter. Temperature: 60°F in summer, 50°F in winter.

Species

C. hookeriana – orange and vermilion flowers
C. ignea, the true cigar flower – black, white and scarlet flowers; evergreen. There is a white variety, but it is rather uninteresting

DATURA (Trumpet Flower)
Glorious long scented white trumpets in August.

Cultivation. Propagate in April or September by 6 in. long cuttings inserted in sandy peat at 65°F. Pot in Levington or Alexpeat compost No. 3 and re-pot into 10 in. pots in March. As soon as flower buds are seen feed with Marinure once a week for two months. In warm years it is possible to stand the plants out of doors in July and August. They require a temperature of 60°F from April to the end of August, and 50°F for the rest of the year. It is wise to prune the branches back by half each year.

Species

D. suaveolens – 8 ft. tall
D. cornigera – white or cream funnel-shaped flowers
D. sanguinea – 4–6 ft. tall, flowers orange-red and 8 in. long

DIONAEA (Venus' Fly Trap)
This is an insectivorous plant, trapping flies between its pairs of lobed leaves, and digesting them by special juices. Produces flowers in July and August. Can be raised from seed sown in March or April on peat.

Cultivation. Divide the plants in March, and pot in well-drained pots of equal parts sedge peat and living sphagnum moss. Re-pot in April. The pots should be stood in a dish or pan containing about 1 in. of water. Give plenty of water at all times, and keep the greenhouse atmosphere moist. When grown in a sunny spot the leaves become almost red, but remain green if grown in the shade. Temperature: 45°F from April to the end of September, 40°F for the rest of the year.

Species

D. muscipula – white flowers 1 in. across, borne on stems 5 in. long

ERANTHENUM (Lovely Flower)
Worth growing for its ornamental leaves and lovely flowers from April to August.

Cultivation. Propagate in April or May by striking 2 in. long cuttings in the propagating frame at 75°F. Pot in Levington or Alexpeat compost No. 2 and re-pot in March or April. Grow the plants on the greenhouse staging from September to June, at 60°F until March, then 70°F until June, and put in the cold frame from June to September. When the plants are in flower feed with liquid manure once a fortnight; water plentifully in summer and moderately in winter. Immediately after flowering, prune the shoots back to within 1 in. of their bases.

Species

E. macrophyllum – pale blue flowers 1½ in. long
E. pulchellum – dark blue flowers 1¼ in. long
E. roseum – rose coloured flowers 1½ in. long

ERICA (Heath)
Flowering from December to August.
Cultivation. Propagate in summer by ½ in. long cuttings of old wood, in the propagating frame. Pot in Levington or Alexpeat compost No. 1. re-potting very firmly in April. Ericas should never be over-watered, in fact it is better to keep them on the dry side, but they should be fed with Marinure once a fortnight in summer. Because the plants hate lime, try to use rain water for watering. They require a temperature of 50°F and need all the sunlight possible in winter. Pinch back loose shoots so as to keep the plant compact in shape. The hard-wooded types need little pruning, but the soft-wooded ones should be pruned after flowering.

Species

E. cavendishiana – rich yellow flowers, May to July
E. gracilis – pink flowers, September to December
E. hyemalis – a soft-wooded kind with rose-tinted tubular flowers, December to March
E. pageana – rich yellow bell-shaped flowers, March–April
E. ventricosa – pink flowers, June. There are also rosy-red, rose-purple and white varieties

ERYTHRINA (Coral Tree)
Deciduous shrub bearing scarlet pea-shaped flowers from May to June.

Cultivation. Propagate in April by taking 2 in. long cuttings of young shoots with a small portion of the old wood attached and insert in sandy peat in the propagating frame at 75°F. Pot in Levington or Alexpeat compost No. 3 and re-pot in March. Place in a sunny position on the staging in March. Water freely from May to the end of September, and keep almost dry in winter and spring. After flowering the shoots die and the pots may be laid on their sides under the staging during winter. Temperature: 45°F October to March, 55°F from April to

September. Prune the shoots hard back to within 1 in. of the old wood in October

Species

E. cristagalli – prickly stalks and deep scarlet flowers
E. fulgens – narrow brilliant scarlet flowers 1½ in. across

EUPATORIUM (Hemp Agrimony, Joe-Pye Weed)
A Mexican plant flowering in the autumn.

Cultivation. Propagate in April by 2 in. long cuttings inserted in sandy peat soil in the propagating frame. Pot up into 3 in. pots of Levington or Alexpeat compost No. 3 and re-pot into 8 in. pots. The pots can go into the cold frame throughout summer, and there the leading shoots should be pinched back every five weeks or so to keep the plants bushy. Temperature: 50–55°F. Feed with Marinure once a week during growth (if the leaves turn yellow it is an indication that they need food), water freely during summer, moderately in winter and spring.

Species

E. atrorubens – reddish flowers with lilac shading
E. micranthum – white flowers, sometimes tinged with rose
E. ianthinum – large purple flower-heads

EUPHORBIA
E. splendens and *E. fulgens* flower in summer and autumn, *E. pulcherrima*, the Poinsettia, in autumn and winter. The 'flowers' of the Poinsettia are in fact large coloured bracts surrounding tiny flowers, and not petals, and technically it should be classed as a foliage plant.

Cultivation. Propagate by striking cuttings of young shoots in the propagating frame (the bases of the cuttings should be dipped in charcoal first). Pot in Levington or Alexpeat compost No. 3, re-pot in March or June and grow in a sunny position in the greenhouse at 55–60°F. Keep on the dry side all through the year, and also keep the atmosphere dry. Cut back the shoots of *E. fulgens* quite severely in June.

Cultivation of Poinsettia. Propagate by 2 in. long young shoots in the propagator at 85°F. Pot in Levington or Alexpeat compost No. 3, re-potting in March or June. When the plants are growing well, water freely, syringe twice daily and feed twice a week with liquid manure during the summer. Temperature: spring 50°F, summer 65°F, winter 55°F. After flowering, lay the pots on their sides beneath the greenhouse staging.

Species

- *E. splendens*, Crown of Thorns – very long prickles and tiny crimson flowers. See Fig. 26, p. 210
- *E. fulgens*, Scarlet Plume – 2–3 ft. tall, can be trained in a circle. Scarlet flowers
- *E. pulcherrima*, Poinsettia – 3–6 ft. tall, gorgeous scarlet bracts

FELICIA (Blue Marguerite, Cape Aster)
A small shrubby herbaceous perennial flowering in June, July and August.

Cultivation. Propagate in July or August by cuttings of young shoots struck in the propagating frame at 60°F. Pot in Levington or No-Soil compost No. 2, re-pot in May and grow at a temperature of 50°F. Pinch back the growing points three times, at intervals of three weeks, to encourage bushy growth. Specimens required to flower in late summer should first go into a cold frame in early summer.

Species

- *E. coelestias* – blue flowers on a 15 in. stem
- *E. petiolata* – prostrate, with rose-coloured flowers

FRANCOA (Bridal Wreath)
There is some argument as to which of the two species below is the true Bridal Wreath; I think it is *F. ramosa* – the white.

Cultivation. Pot in Levington or Alexpeat compost No. 2, re-potting in March or April and dividing the plants at this time. Propagation can also be by seed sown in sandy peat at 50°F in March or April. When the plant is well rooted, feed once a week and give plenty of water in summer and little in winter. Temperature: 60°F April to September, 50°F October to March. The plants are normally thrown away after four years.

Species

- *F. ramosa* – lovely white flowers on 2–3 ft. stems
- *F. sonchifolia* – loose racemes of pink flowers with darker spots near the bases

FUCHSIA

Fuchsias are increasing in popularity, and can be grown as small bushy plants or as standards, both flowering all summer. Numerous beautiful varieties are available from those nurserymen who specialise in fuchsias.

Cultivation. Propagate by cuttings or seed, using Levington or Alexpeat compost No. 2 and re-potting in February or March. Grow the plants in the shadier part of the greenhouse from March to July, and in the sun from July to October at 55°F. From October to March put the plants beneath the staging at a temperature of 40°F. Water moderately in spring, freely from May to mid-October, and give liquid manure once a week in summer from the time the flowers are seen. Syringe the plants over daily from mid-February to the end of May. Prune the branches of old plants back by half each February.

Varieties I have grown are:

Lena – large double mauve and white
Mauve Beauty – large mauve and red
Mrs Popple – red and purple
Tom Thumb – dwarf red and purple
Peggy King – red and blue
Susan Travis – dwarf pink and purple

HEDYCHIUM (Ginger Lily)

A very easy and lovely plant to grow, flowering in July and August. Not suitable for the small greenhouse as it reaches a height of 5 ft.

Cultivation. Pot in Levington or Alexpeat compost No. 3 and re-pot into 8 in. pots in March or April, dividing the plants at this time. The March to November temperature should be 55°F, and 45°F for the rest of the year. Water well in summer, feeding twice a week with Marinure once the plants are in flower, but only occasionally in winter. Cut flowering stems down immediately flowering is over.

Species

H. flavum – dense spikes of yellow and orange flowers
H. gardnerianum – lemon tube-like flowers with wedge-shaped lips
H. spicatum – loose spikes of long yellow flowers in October

HELIOTROPIUM (Heliotrope, Cherry Pie)

A lovely scented plant, flowering from June to September.

Cultivation. Propagate in July by striking cuttings in the propagating frame. Pot up soon afterwards into 3 in. pots of Levington or Alexpeat compost No. 2 and overwinter in a frame or cool greenhouse. Re-pot into 6 in. pots in spring and grow at 60°F from February to October, 50°F from October to February. When plants are growing well, feed weekly with Marinure. Pinch out the growing points in February or March to keep the plants bushy. Old plants need pruning quite hard every winter, and should be re-potted into new compost each spring.

Species

H. peruvianum – violet or lilac flowers, very fragrant
H. anchusifolium – scentless mauve flowers in May

HUMEA (Amaranth Feathers)
An Australian plant with a lovely 'cedar-wood' scent, flowering from June to October.

Cultivation. Propagate by seed sown in Levington compost at 60°F in April, May or June. Pot on into 3 in. pots of Levington or Alexpeat compost No. 3 when 1 in. tall, and re-pot in April. Some grow to 6 ft. and should eventually be potted into 8 in. pots. When the flower spikes appear, feed with Marinure once a week. Water carefully, especially in winter, as they quickly die off if over-watered. Throw away the plants after flowering. Temperature: 55°F April to October, 45°F for the rest of the year.

Species

H. elegans – pinkish or crimson flowers in feathery panicles

HYDRANGEA
In bloom from July to September.

Cultivation. Propagate in May, June or July by 3 in. long cuttings struck in peaty sand in the propagating frame. Dip the bases of the cuttings in a hormone dust if rooting difficulties are experienced. The moment rooting begins, give a little air to the cuttings, and then pot into 3 in. pots of Levington or Alexpeat compost No. 2 and stand plants in a frame. Two or three weeks later pinch out the growing points to encourage bushy growth. Pot up into 6 in. pots when the plants start to get pot-bound, the best time to re-pot being March. Keep in a frost-proof frame or cool greenhouse during winter, and give very little water or the leaves will fall off. In mid-February bring on to the

1. *Callistemon citrinus* 'Splendens', p. 100
2. *Clerodendrum thomsonae* var. *balfouri*, p. 17
3. *Codiaeum variegatum pictum*, p. 186
4. *Coleus blumei*,
5. *Cordyline terminalis* (*Dracaena*), p. 187
6. *Cyperus alternifolius*, p. 187

Hydrangea

Pelargonium

Cypripedium Orchid

FIG. 23 Three typical pot plants

greenhouse staging at 50°F and water well. Grow on at 55–60°F. Continue watering and syringing over and the plants will soon break into leaf. A solution of alum artificially changes pink-flowered varieties to blue. Soil must be lime-free.

Species and varieties. H. macrophylla is the common hydrangea, bearing lovely pink, white or blue flowers 6 in. across. The variety Nigra has purple stems and bright rose flowers, and nurserymen's catalogues feature numerous other unusual varieties.

IMPATIENS (Busy Lizzie, Balsam)

A very easy plant to grow, in flower almost the whole year round.

Cultivation. Propagate by seed in spring or by 1 in. long cuttings struck in the propagating frame in summer. Potting compost should be Levington or Alexpeat No. 1 and re-potting may be done almost any time. Water regularly to keep the compost just damp, and feed with Marinure once a week as soon as flowering begins. Prune back the plants from time to time to keep them bushy. Temperature: 55°F.

Species

I. balsamina – 20–24 in. tall, with brightly coloured flowers of scarlet, violet, white or salmon-pink. The camellia-flowered strain is particularly beautiful, and it is also worth trying the Tom Thumb mixed

IXORA

Very fragrant blooms in summer, sometimes into October.

Cultivation. Propagate in April by 2 in. long cuttings of firm shoots, and strike in sandy peat in the propagating frame at 75°F. Pot in Levington or Alexpeat compost No. 2 and re-pot in February or March, always using plenty of crocks as good drainage is important. Water freely in spring and summer, moderately in autumn and winter. Temperature: 75°F March to September, 55°F October to February. Syringe once a day (but preferably twice) from early April to the end of August. Every year, prune the plants into the shape required.

Species

I. chinensis – many-flowered and much-branched, with bright orange blooms. There are white and pink varieties, and a cinnabar-red one called Prince of Orange
I. coccinea – bright red flowers, often with tubes 2 in. long. The variety Pilgrimii has bright orange-scarlet flowers
I. congesta – bright orange-red flowers

JACOBINIA

Flowering in August and September, one variety even flowering in December.

Cultivation. Propagate in April, May or June by striking 2 in. long cuttings of young shoots in sandy peat in the propagating frame at 75°F. Pot in Levington or Alexpeat compost No. 2 and re-pot in March or April. Keep the plants in the greenhouse from September until June, and then put them in a frame until September again. Give plenty of water from March to September, little from September to March, and feed with Marinure twice a week when the flower buds are seen and until flowering is over. Temperature: 55°F from September to March, 65°F March to June. Pinch out growing points in June and July to encourage bushy growth and cut back stems after flowering to within 1 in. of their bases. Red spider can be a serious pest, and should be controlled by frequent syringing over of the plants.

Species

J. aurea – yellow tubular flowers, July
J. carnea – fresh-coloured flowers, August and September
J. coccinea – groups of scarlet flowers
J. pohliana – bright crimson flowers
J. p. velutina – pink flowers

JASMINUM (Jasmine)

These attractive winter-flowering plants can be trained to wires near the roof, and left there permanently.

Cultivation. Propagate any time from April to August by striking 2 in. long cuttings of firm shoots in the propagating frame at 70°F. Pot in Levington or Alexpeat compost No. 2 and re-pot in February or March. Grow at 55–60°F and give plenty of ventilation. Water freely in spring and summer and moderately for the rest of the year. Avoid over-potting as this causes a reduction in flowering. These plants do well in hanging baskets in full sun.

Species

J. mesnyi – scented yellow flowers
J. officinale grandiflorum – white flowers in autumn

KALANCHOE

A succulent flowering in May and June and sometimes scented. Likes full sun.

Cultivation. Propagate by 2 in. long cuttings in July or August. Let the cuttings dry for two days before inserting them into the sand in the propagating frame. Pot in Levington or Alexpeat compost No. 2 and re-pot in July or August. Alternatively, grow from seed sown in Levington compost in March, potting up the seedlings into 6 in. pots in July. Water well from April to the end of August but moderately from September to November. Temperature: 60°F April to September, 50°F for the rest of the year. Prune the stems after flowering to within 2 in. of their bases, re-potting when the stems start to grow again.

Species

K. blossfeldiana – 1 ft. tall, scarlet flowers
K. vivid – bright scarlet
K. thyrsiflora – a tight spike of yellow flowers

LOTUS

A relative of bird's-foot trefoil, flowering in summer.

Cultivation. Propagate by striking 2 in. long cuttings in the propagating frame in summer or by sowing seeds in April at 65°F. Pot in Levington or Alexpeat compost No. 2 and re-pot in March. Grow the plants in full sun on the greenhouse staging, at 60°F from April to September, giving plenty of air at this time, and 50°F from October to March. Water moderately in summer and sparingly in winter, and feed once a week as soon as flowering starts.

Species

L. bertholetii – 2 ft. tall, scarlet flowers

MARGUERITE

Included here under its common name rather than with the other chrysanthemum species in Chapter 12. White flowers in summer.

Cultivation. Propagate by 2 in. long cuttings in April. When the plants are 6 in. tall, pinch out the growing points, and repeat when the side growths have developed. Pot in Levington or Alexpeat compost No. 2 and grow at a temperature of 50–55°F. Water well during the whole of the growing period and feed once a week when well established. The plants can be grown outside in July, August and early September.

Species

Chrysanthemum frutescens – typical daisy-like flowers with yellow centres and white petals

MIMULUS (Musk, Monkey Flower)

Perennials with tubular flowers in summer.

Cultivation. Pot in Levington or Alexpeat compost No. 2 and re-pot in February, March or April. Propagate by division in March, by 2 in. long cuttings struck in the propagating frame at 55°F, or by seeds sown in March, April or May in No-Soil compost at 55°F. Grow plants in the shadier part of the house, at 50–55°F in summer, 45°F in other months. Give plenty of water in summer and feed weekly in spring and summer with Marinure. Pinch back the side shoots to make the plants bushier.

Species

M. glutinosus – a shrubby plant with deep yellow flowers
M. cardinalis, the Cardinal's Monkey – scarlet
M. ringens – 2–4 ft., violet flowers

MOSCHOSMA

A nettle-leaved perennial (also known as Iboza) with white and purple flowers in autumn and winter.

FOLIAGE PLANTS I. *Above left, Cissus discolor; right, Dracaena deremensis* var. *bausei. Below,* Basket-grass, *Ophismenus hirtellus variegatus.*

FOLIAGE PLANTS II. *Maranta leuconeura. Above, erythrophylla. Below, kerchoveana.*

Cultivation. Propagate in March by striking 2 in. long cuttings of young shoots in the propagating frame at 65°F. Pot in Levington or Alexpeat compost No. 2 and re-pot in March and July. After the July potting feed once a week with Marinure. Water well in spring and summer and moderately the rest of the year. Temperature: 60°F for April, May and June, 50°F from August to the end of March. The plants may be put into a cold frame from June to the end of September, but must be in the greenhouse for the rest of the year. Prune back after flowering to within 3 in. of the base of the plant.

Species

M. riparia – 2 ft. tall, white and purple flowers

NERIUM (Oleander, Rose Bay)
Shrubby decorative plants which are easy to grow, flowering all summer.

Cultivation. Strike 2 in. long cuttings taken from the tips of the shoots in pure silver sand in the propagating frame, which should be kept well watered. Pot in 6 in. pots of Levington or Alexpeat compost No. 2 and re-pot in March. Grow in a sunny position as shade makes the stems weak, and syringe daily through summer. Pinch back young growths in summer, and prune in winter to keep the plants bushy.

Species

N. oleander – rosy-pink flowers. There are also double red and white varieties

PAVONIA
Evergreen shrubby plants flowering in autumn.

Cultivation. Propagate by striking cuttings in the propagating frame at 65°F in spring and summer. Pot in Levington or Alexpeat compost No. 2 and re-pot in March. Give plenty of water in spring and summer and daily syringing in summer, but little water in winter. Grow at 45–50°F.

Species

P. intermedia – white flowers
P. multiflora – purple flowers on 18 in. stems

PELARGONIUMS
There are several types, some suitable for hanging baskets, with green or variegated leaves and flowers of different colours.

GERANIUMS

Extremely popular plants flowering from spring to autumn.

Cultivation. Propagate in August or September by striking cuttings in boxes of sandy peat on the greenhouse staging or inserted round the edge of 6 in. pots, at 45°F. When rooted, pot up the cuttings into 3 in. pots using Levington or Alexpeat compost No. 3. Remove any flower buds which may be present, and water moderately for 14 days and normally thereafter. Feed with liquid manure every 14 days when well established, and grow at 50°F. Shade from the sun when flowering. Cut back plants after flowering and re-pot in March. For winter flowering, take cuttings in March.

Varieties

Irene – very large semi-double vermilion
Judith – deep lilac, dwarf. Very prolific
Penny – semi-double rosy-mauve
Red Landry – semi-double salmon-red
Toyon – dwarf, large red flowers
Also numerous other varieties such as Beatrix, Peach, Vestal Blue Peter, La France, Marquis and Mauve Beauty

IVY-LEAVED GERANIUMS

Excellent for hanging baskets, flowering in summer.

Cultivation. Propagate in August or September by striking 2 in. cuttings in the propagating frame. Pot up after striking in Levington or Alexpeat compost No. 2 and put on the greenhouse staging in a slightly heated frame. Pot up into 6 in. pots in March and a fortnight later pinch out the growing points. Water moderately in autumn and winter, growing at a temperature of 40°F, and plentifully in spring and summer, at 50°F. Feed weekly with liquid manure from May to September. Prune old plants back each February, and re-pot in April or May.

Varieties

Blue Peter – magenta and purple
Enchantress – superb rose colour
Galilee – bright shell-pink, double
Marquis – double red, a strong grower

SHOW PELARGONIUMS

Also called regal pelargoniums; flowering in spring and summer.

Cultivation. Propagate in July or August by 2 in. long cuttings. When rooted, pot up into well-drained 3 in. pots of Levington or Alexpeat compost No. 3. When these pots are full of roots, pot on again into 6 in. pots and a week later pinch out the growing points. Place near the glass and then move to the greenhouse staging in September. When in flower, feed once a week with Marinure. Temperature: 45°F September to March, 50°F March to May. Water freely from March to June. Re-pot in January. Prune shoots fairly hard each July.

Varieties

All My Love – orchid-mauve on a cream base
Blush Queen – mauve and purple with white blotches
Grandma Fischer – salmon-orange with a bronze blotch
Grand Slam – deep rose, shading to crimson
Rogue – mahogany and crimson, with darker markings

PRIMULA

Rosettes of leaves with single or double flowers in winter and spring.

Cultivation. Propagate by seed sown in March and May for *P. obconica*, *P. malacoides* and *P. sinensis*; January and February for *P. kewensis* (which needs a longer season of growth), and April for *P. auricula*. The plants may also be divided after flowering. Pot up lightly into 3 in. pots of Levington or Alexpeat compost No. 2 (No. 1 for auriculas) and into 6 in. pots when the roots have filled the 3 in. ones. (Auriculas should be left in the 3 in. pots.) Primulas like plenty of ventilation, a temperature of 50–55°F and they should be grown in the shadier part of the greenhouse or frame. The compost should be kept just moist, as over-watering causes yellowing of the leaves. *P. auricula* should be watered moderately in winter but freely in spring and summer. Feed with Marinure once a week when the plants are growing well and the auriculas are in flower. Re-potting may be done almost any time, but in February or March for auriculas.

Species

P. obconica – large flowers in shades of pink and mauve. The leaves cause eczema in some people
P. malacoides – tall candelabra-like flowers in pink, rose, lilac and white
P. kewensis – yellow flowers, a little smaller than *obconica*
P. sinensis – large frilled flowers in rose, scarlet, crimson, mauve, pink and white
P. auricula Alpine Mixture – many differing kinds of dwarf plants
P. auricula Douglas Prize Mixture – tall, beautiful colours

REHMANNIA

Really a perennial, but usually treated as a biennial. Flowers somewhat like foxgloves in spring and summer.

Cultivation. Sow seed early in May in Levington compost. Transplant the seedlings into 3 in. pots of Levington or Alexpeat compost No. 2, and place on the staging at 50°F. In autumn, pot on into 6 in. pots at 45°F, raising the temperature slightly again to re-start growth the following March. Re-pot in February or March. Cut out the first flowering stem so as to encourage the production of flowering side shoots. Feed once a week with liquid manure when the pots become filled with roots. Spray with Malathion to prevent greenfly and white fly.

Species and varieties

R. angulata – 1–2 ft. tall, red and orange flowers
R. alba – white flowers
R. Pink Perfection – pink flowers

REINWARDTIA (Yellow Flax)

An evergreen shrub flowering in autumn and winter.

Cultivation. Propagate in April by 2 in. long cuttings inserted in sandy peat in the propagating frame at 65°F. Pot in Levington or Alexpeat compost No. 2 and re-pot in March or April. Grow plants on the staging in the sun at 45°F, and put them outside in the summer, if desired. Feed with Marinure once a week during the growing season, and syringe twice daily from March to the end of September. Pinch out the growing points of young shoots in June to promote bushiness, and prune back hard in February.

Species

R. indica – 2–3 ft., yellow flowers in autumn
R. tetragyna – yellow flowers, winter

ROCHEA (Kalosanthes)

A succulent flowering in spring and summer.

Cultivation. Propagate by 2 in. long cuttings taken from the tips of non-flowering stems when the plants are being dried off. Strike in the propagating frame at 65°F and do not apply water while the cuttings are trying to make roots. After rooting, pot up into 3 in. pots of Levington or Alexpeat compost No. 1, plus a little mortar rubble. Give a fair amount of water during summer, and a temperature of 50°F, but keep almost dry in winter, at 45°F. Grow the plants in sunshine, especially when flowering is over, as new growths need to be thoroughly ripened. Pinch back the growing points for more bushy growth. After flowering, prune back old plants to within 1 in. of their bases. Re-pot in March.

Species

R. coccinea – beautiful scarlet flowers on 1 ft. stems
R. versicolor – white and pink flowers, spring

ROSA (Roses) – see p. 123.

RUELLIA (Christmas Pride)
Largely Brazilian plants with funnel-shaped flowers in winter.

Cultivation. Take cuttings in spring or summer and strike them in sandy peat in the propagating frame at 75°F. Seeds of some varieties may be sown in March at 70°F, in boxes of Levington compost on the greenhouse staging. Pot into Levington or Alexpeat compost No. 2, re-potting in February or March. Grow the plants in the shadier part of the greenhouse and give lots of water from the start of April to the end of September, but little after this time. Syringe over twice a day in spring and summer, once every two days in autumn and winter. Feed with Farmura once a week during the flowering period. Temperature: 55°F September to March, 65°F for the rest of the year.

Species

R. macrantha, the true Christmas Pride – rosy-purple flowers in winter
R. portellae – rose-pink flowers in winter
R. solitaria – purple flowers in winter

SAINTPAULIA (African Violet)
A close relative of the gloxinias, with fleshy hairy leaves and normally deep violet flowers in autumn and early winter. Can be raised from seed.

Cultivation. Propagate by means of leaf and petiole cuttings pushed into silver sand in the propagating frame at 60°F. When young growth is visible at the top of the leaf, pot up into 3 in. pots of Levington or Alexpeat compost No. 2. Re-potting may be done any time from February to May. Grow in a position shaded from strong sunlight at a steady summer temperature of 60°F and a winter one of 50°F. Never water over the leaves, but give moisture by immersing the pots in water. Marinure may occasionally be added to the immersion water during the flowering season.

Species

S. ionantha grandiflora, Wicks Gold Medal hybrids – a new strain with blue, purple, pink, white and near-red flowers

SALVIA (Flowering Sage)

Easy to grow perennials flowering in autumn and early winter and suitable for a small house. Can be raised from seed.

Cultivation. Take 3 in. cuttings in March and strike in the propagating frame at 65°F, or sow seed in March in Levington compost at 60°F. Pot in Levington or Alexpeat compost No. 1 and grow the young plants in a cold frame from June to September, if desired, but they must come into the greenhouse at 45°F in early October. Water well in spring and summer, but very carefully in winter. The young plants must be potted on as soon as possible, and the growing points pinched out to encourage bushy growth. When flowering is over, prune back the shoots to within 3 in. of their bases.

Species and varieties

S. splendens, Blaze of Fire – vivid bright scarlet flowers. 2–3 ft. Good varieties are Fire Ball, Gypsy Rose (dusky-rose), Purple Blaze (reddish-purple), Salmon Pygmy (dwarf salmon-red)
S. patens – the only blue species. 2–3 ft.

SARRACENIA (Pitcher Plant)

An insectivorous plant with pitcher-shaped flowers in spring and summer.

Cultivation. Grow in well-drained pots containing equal parts sedge peat and chopped sphagnum moss. Re-pot in March. Keep shaded from bright sunlight in a cool house (45–50°F) with a moist atmosphere. Water freely in spring and summer but very little in winter. Syringe over daily in summer and add a top dressing of chopped sphagnum moss as a mulch.

Species

S. flava, Huntsman's Horn – yellow flowers
S. purpurea, Huntsman's Cup – purple flowers
S. rubra, Indian Cup – reddish flowers
Nurserymen may also recommend a number of hybrids

SENECIO

In some catalogues cineraria may appear as *Cineraria cruenta* or as *S. cruentis*. It flowers in winter or spring, according to when the seed is sown.

Cultivation. Sow seeds in Levington compost in May or June for winter flowering and late June for spring flowering. Transplant the seedlings when 2 in. tall into 3 in. pots of Levington or Alexpeat compost No. 2 and put in the cold frame. Pot up into 6 in. pots in August, and re-pot as necessary. Bring into the greenhouse in early October and place on shelving near the glass at 50°F. Feed once a week with Marinure from early October until in full flower. Water very carefully all the time.

Species

Several seedsmen's strains are particularly good, e.g. Berlin Market, the Hansa strain, and the coppery Scarlet Matador

SINNINGIA (Gloxinia)

Tuberous-rooted plants with large red, pink, violet or blue velvety flowers in autumn.

Cultivation. Sow seed in February in Levington compost at 75°F. When plants are 1 in. tall, pot up into 3 in. pots of Levington or Alexpeat compost No. 3 and two months later pot on into 6 in. pots. Alternatively, cuttings 2 in. long can be taken in spring and rooted in the propagating frame; no water must be given, however, until growth begins. Leaf cuttings may also be taken, as advised for Streptocarpus. Tubers of 1-year-old plants may be rooted in 3 in. pots and later potted up into 6 in. ones. Grow on the greenhouse staging at 65–70°F with heavy shading. Marinure may be applied once a week until the flowers appear, but then both water and food must be reduced to a minimum. Store tubers in sand at 50°F over winter to prevent shrivelling.

Species

S. speciosa (syn. *Gloxinia speciosa*) – violet tubular flowers in autumn

Seed of Multiflora Double-flowered strains, Blackmore and Langdon's Mixed are worth trying, also the slipper-type hybrids

SOLANUM (Winter Cherry)

Relatives of the potato, tomato and deadly nightshade, bearing beautiful orange berries in winter.

Cultivation . Sow seed in February in Levington compost at 65°F. Prick off the seedlings into 3 in. pots of Levington or Alexpeat compost No. 2 and place on the greenhouse staging at 65°F. During April and May reduce the temperature to 55°F and give more ventilation. Spray over daily with water. When 3 in. tall, pinch out the terminal buds to cause the laterals to grow, and put up into 6 in. pots before the end of May. A week later, pot into the cold frame and give each plant a pinch of fish manure. Feed every two weeks from then on with a liquid fertiliser, and syringe frequently in warm weather. Bring into the greenhouse at the end of September, and continue the feeding and syringing. Re-pot at the end of May.

Species and varieties

S. capsicastrum – red cherry-like fruit
Covent Garden – orange and scarlet berries
Patersoni – a very compact shape

SONERILA

Green and white ornamental foliage and rose-coloured flowers in summer.

Cultivation. Propagate by cuttings in February, March or April at 75°F in the propagator, or by sowing seed shallowly in Levington compost at 75°F during the same months. Shade the plants from strong sunlight and grow in well-drained pots of Levington or Alexpeat compost No. 2 at 70°F from March to September, and 55°F for the rest of the year. Water freely in summer, moderately in autumn and winter, syringing over every day in summer as the plants like a moist atmosphere. Re-pot in February or March.

Species and varieties

S. argentea – lilac-rose flowers
S. margaritacea – pink flowers, leaves white and green on top and purplish beneath
Varieties such as Victoria and F. Marchand may be offered by nurserymen

SPIRONEMA

A rather curious Mexican plant with large leaves and white flowers in May.

Cultivation. Propagate by 2 in. long cuttings in the propagating frame

– they strike easily. Pot in Levington or Alexpeat compost No. 1 and re-pot only when necessary. Prefers to grow as a hanging plant in a shady spot at about 65–70°F, and should be fed with liquid manure once a month in summer. Water freely in summer, very little in winter. Can be used in hanging baskets outside during summer.

Species

S. fragrans – small white scented flowers; the only species in the genus

STRELITZIA (Bird of Paradise Flower)
A handsome plant, usually about 4 ft. tall, and rather large for the small greenhouse. The beautiful flowers, green, orange and blue, and shaped like a bird's head, come out in spring.

Cultivation. Propagate in spring by division, by suckers or by seed sown at 65°F in a propagating frame with bottom heat. Pot in Levington or Alexpeat compost No. 3, eventually growing the plants in 8 in. pots in full sun. Temperature: 65°F in spring and summer, 55°F in autumn and winter. Water well in summer, sparingly in winter. Re-pot in February or March.

Species

S. reginae, the true Bird of Paradise flower – green, orange and blue flowers
S. augusta – up to 18 ft. tall, white and purple flowers

STREPTOCARPUS (Cape Primrose)
Produces masses of bell-like flowers in varying shades of pink, red, violet and white in spring, summer and autumn.

Cultivation. Propagate by seed sown shallowly in Alex or Levington compost in January or February. When a good cluster of roots has formed, pot into 3 in. pots of Alexpeat compost No. 2. In six or seven weeks' time pot on into 6 in. pots of Alexpeat compost No. 3. Alternatively, propagate by leaf cuttings. Choose mature leaves, remove the bottom part of the stalk, sever the leaf's main vein transversely in three places and peg it down with hairpins in sandy peat in the propagating frame at 60°F. Small plants will appear where the vein was severed, and these should be potted up into 3 in. pots. Do not allow the leaves of the plant to touch one another. Grow on the greenhouse staging at 55°F and feed with Marinure once a fortnight in summer. Water well in summer but keep on the dry side in autumn and winter. Re-pot, always lightly, in March or April.

Species

S. dunnii – rose flowers, summer
S. galpinii – mauve and white flowers, October
S. wendlandii – violet-blue and white flowers, spring
Many hybrids are also offered by nurserymen

TRACHELIUM (Blue Throat Wort)
A half-hardy herbaceous perennial which can be raised in pots annually from seed. Dense heads of tiny flowers in July and August.

Cultivation. Propagate in spring or July by seed sown in Levington compost at 55°F. Pot in Levington or Alexpeat compost No. 2, and re-pot in March or April. Pinch back the growing points of the young plants to produce bushy growth. The plants resulting from a July sowing should be put in a cold frame until early October, when they can be put on the greenhouse staging at 45°F. Feed with Marinure once a week during summer, and grow in full sun.

Species

T. caeruleum – 2 ft., beautiful blue flowers, August
T. album is a white variety

VRIESIA
A Brazilian plant very similar to Tillandsia. Flowers in summer.

Cultivation. Propagate in April by off-sets potted up into 3 in. pots of Levington or Alexpeat compost No. 2. Give plenty of water from April to mid-October, and little from then on. Grow in the shade and syringe over daily from April to the end of September maintaining a moist atmosphere. The temperature for this period should be 70°F, and 60°F from the beginning of October to the end of March. Re-pot in March.

Species and varieties

V. psittacina – yellow flowers spotted green at the tips, July. Morreninia is a good variety, with numerous closely-set flowers
V. splendens – erect yellow flowers, leaves with dark brown bands and red bracts
V. tessellata – yellowish-green flowers 2 in. long
V. t. roseo-picta – has large rosy spots
V. hieroglyphica – dull yellow flowers with dark green leaves, blackish-purple below

ZANTEDESCHIA

A beautiful, not very hardy, perennial with white flowers from Christmas to Easter.

Cultivation. Propagate in August by off-sets or division of crowns in a compost consisting of 6 parts medium loam and 1 part medium-grade sedge peat. An egg-cupful of hoof and horn meal should be added per 8 in. pot of the compost. Put the off-sets into 3 in. pots; they take about three years before they are ready to flower. After potting, stand the plants outside on ashes and water regularly. At the end of September bring them into the greenhouse at 60°F, continue regular watering and feed with Marinure once a fortnight. After flowering, rest the crowns, watering moderately. Place the pots outside in May and lay them on their sides early in June. The corms will gradually die off and be ready for re-potting in August.

Species

Z. aethiopica, White Arum Lily – typical white lily-like flowers up to 10 in. long

Z. elliottiana – beautiful yellow flowers in August. With this species, the crown must be re-potted in February and the temperature should be 70°F

ZINGIBER (Ginger Plant)

A perennial plant with large leaves and spikes of flowers in July. It is from the tuberous roots that stem and crystallised ginger are made.

Cultivation. Propagate by division of the roots in February. Pot in Levington or Alexpeat compost No. 2, re-potting in February. Water copiously in spring and summer, but keep almost dry in autumn and winter. Grow the plants at 75°F from April to September and 55°F from October to March, in a shady part of the greenhouse. Syringe over regularly as these plants love moisture. Let the stems die down after flowering.

Species

Z. officinale – greeny-yellow and purple flowers, July

ROSA (Rose)

Those who want to grow roses in pots in the greenhouse must take up the matter seriously; it isn't just a question of digging up an old hybrid tea or floribunda from the garden and potting it up. Roses in the greenhouse are not necessarily easy to grow, but if the instructions given in the notes below are followed, the reader will have a glorious spring display in his greenhouse.

Propagation. This is done either by budding or grafting, i.e. by taking buds, or small pieces of well-ripened shoots of the desired variety and making these grow on a suitable stock (a 'briar' will do), which must be well rooted and growing steadily.

1. ROSES TO BE GROWN FOR CUT FLOWERS

The selected rose-bushes may be potted up in February in 3½ in. pots using Levington compost. The pots should be stood outside in a cold frame, plunging them deeply in sedge peat so as to prevent excessive drying out.

They should be potted up again in late May or early June using the following compost:

3 parts good fibrous loam
1 part well-rotted compost or old mushroom bed
1 6 in. pot of bone-meal per barrow load of compost
1 handful of dry wood ash per barrow

The pots should be well crocked and the compost fairly coarse. Potting should be very firm, but 1 in. should be left at the top to allow for watering. Plants should be stood outside as soon as the potting is completed and any leaves should be removed; by mid-August they should have made nice bushy plants.

The plants may be brought into the house in batches of two or three at fortnightly intervals from early December until early January, allowing sufficient room between the pots for growth. Full ventilation should be given – frost will not harm them and they may, in fact, be allowed to become very dry as this will help to ripen the wood properly. During this period no heat should be given.

Pruning should be done during the last two weeks of December for the early batches. The longer this is delayed the later the flowering. When severe pruning is necessary, the plants should be cut back hard to plump buds, removing all weak straggly growths. It is usual to expect six blooms per plant in the first year, and flowering usually takes place in 14 to 16 weeks from pruning.

Temperature and ventilation. Ventilation should be reduced at the end of January, and as soon as growth starts a temperature of 50–60°F should be maintained during the day and about 45°F at night. The temperature may, however, rise to 65–70°F with sun heat. During

March the temperature may be raised to 60–70°F during the day and 60°F at night.

Shading. As soon as the sun heat begins to get really strong shading is necessary, in the form of either whitening or slatted blinds.

Watering and damping. It is important not to give any water till root development begins. Syringing and damping down should be done whenever the weather is favourable, but this must stop when buds show colour. Watering should be slowly increased from January onwards until the plants are taking it every day. This work should be done preferably during the morning.

Feeding. As soon as the plants are brought under glass they should be given a top-dressing of fresh soil plus 'manure' consisting of a mixture of 2 parts Peruvian guano and 3 parts wood ashes, at 3 teaspoonfuls per pot. Feeding should be done generously, preferably every two or three days if possible. Weak liquid manure has also proved successful.

Disbudding. There should be three to four stems only on each plant, so remove all the inside shoots and weak growths. Any lateral flower-buds that appear should be removed also, 'taking' the terminal bud only for flowering. Plants flower for about four to six weeks and will give a succession of flowers from March until May and June. During June the heat is turned off and more ventilation given.

After flowering the bushes should be cut back by half and rested for two weeks or so, giving full ventilation and no syringing. After this a second crop may be produced if the plants are watered and fed as before.

The full resting stage is from September until October. The plants should be re-potted every two years and a top-dressing given every alternate year.

2. ROSES TO BE GROWN AS DECORATIVE POT PLANTS

The procedure is exactly the same as before, except for the fact that it is a good thing to disbud the blooms of the floribundas, apart from the fact that the roses should be potted up into 6 in. or 8 in. pots.

Staking and tying. It is seldom that floribunda roses need any staking or tying, but some taller types may need a central bamboo and a tie-round with raffia.

Varieties. The hybrid tea species are usually considered the best for pot roses as well as for cut blooms. Dwarf floribunda roses, however, may be in greater demand as decorative flowering pot plants. The following are the varieties I have grown with success.

Richmond – a vigorous rose having rich red flowers with a very sweet scent
Baccara – an excellent crimson rose
Madam Butterfly – pink, with pointed buds and a distinct tea scent
Lady Sylvia – deep pink, a sport from the variety Madame Butterfly
Briarcliffe – large and full rose-pink flowers
Ernest Morse – dark red, upright growing, sweetly scented
Mrs Charles Lamplough – a soft lemon-yellow
Royal Highness – soft light pink, fragrant
Doctor Verhage – a chrome yellow, long lasting
Virgo – a pure white
Edel – pure white. Usually grown for exhibition work only
Mrs Foley Hobbs – ivory-white

DWARF FLORIBUNDA

Cameo – salmon-pink. Very pretty
Ellen Poulsen – deep pink
Golden Salmon Superior – golden-salmon
Paul Crampel – orange-scarlet; a brilliant colour
Perle d'Or – yellow shaded peach, tiny double flowers

CHAPTER II

Flowers for Borders, Bedding and Cutting

DAHLIA

The dahlia is not generally considered to be a greenhouse plant, but when grown under glass, however, one can get them to flower early, four or five weeks in advance of the outdoor plants. They have tuberous roots and opposite divided leaves. The height ranges from 18 in. up to 5 or 6 ft. and the flowers may be single or double, occurring in a very wide range of colours including pink, orange, scarlet, crimson, yellow, magenta, white and cream.

Cultivation. The gardener who is intending to produce some early dahlias under glass will probably also have some plants outdoors. Some of the tubers should be potted up in May in 6 in. pots and allowed to grow and flower in these during the summer. The pots should be kept watered through the summer and a temperature of 55°F maintained at night, rising to 60–65°F during the day. The plants must not on any account be forced. From the middle of September onwards water should gradually be withheld, stopping completely during the resting period. As soon as the plants have died down, the tubers should be lifted and stored in a dry frost-proof shed.

In January these pot tubers, as they are often called, should be potted singly in 3 in. pots, using Levington potting compost. Firm potting is essential. The pots should then be placed on the greenhouse staging where they can receive plenty of light and air.

The soil should be kept on the dry side until growth becomes really active, as too much water in the early stages may cause the tubers to rot. Later on, copious supplies will be required, and in hot weather the plants may be syringed over each morning.

Dis-shooting should be done about the beginning of February, when the best and strongest shoots should be selected and the others rubbed out, retaining about six per plant. The plants should be potted on a fortnight or three weeks later into 10 in. or 12 in. pots of Levington potting compost according to the type of dahlia being grown, or planted directly into the greenhouse border.

The taller types will require staking and tying, and this should be done in the early stages as the soft sappy shoots are easily broken.

Feeding. This should begin about a fortnight after the plants have been potted and placed in their flowering positions, after which a weekly feed may be given. Marinure is the ideal feed to use, giving roughly a quart of the diluted liquid per plant at each feed.

Preparation of the borders. The borders should be dug in the autumn or winter, incorporating some well-rotted organic matter, as dahlias are heavy feeders. After digging, and a few weeks before planting out, fish manure or some similar organic fertiliser such as meat and bone-meal or hoof and horn meal should be worked into the surface soil at the rate of 4 oz. per square yard. After dis-shooting, the plants should be set firmly in the prepared border and watered in well. Distances apart will depend on the size of the dahlias being grown; some of the larger species may require up to 3 ft. The plants do not as a rule require any feeding.

Cutting. The blooms should be cut just before they reach maturity, and the ends of the stems immediately plunged into boiling water for a few seconds, after which they should go into cold water for a few hours.

Stock plants. Good, healthy plants should be selected in the summer and set on one side for stock. After flowering, the tops should be cut down to within 3 or 4 in. of ground level, and the roots lifted a few days later. Careful storage is essential if the tubers are to overwinter successfully, and the building where they are to be stored must be dry and frost-proof.

Varieties

Adorable You – lavender-pink, with fine stems
Beauty of Crofton – long cone-like stems, rich salmon-apricot flowers
Bruno Walter – strong salmon-red on good long stems
Corona – brilliant vermilion-scarlet with yellow base and crimson at the centre. A bedding cactus type
Dedham – early-flowering, lilac suffused with white, good stems
Doris Day – outstanding cardinal-red
East Anglian – rich golden-yellow with faint salmon hues
Force Eight – deep crimson-red, with a hint of purple
Gerrie Hoek – silvery-pink, one of the best pinks
Golden Turban – golden-apricot with bright deep yellow centre
Hamari Bride – white medium Cactus
House of Orange – pure orange
Jescot Duo – vermilion-red, tipped white
Jescot Jim – a good yellow

FOLIAGE PLANTS III. *Above, Sonerila margaritacea argentea. Below, Vriesia fenestralis.*

Above, small, decorative dahlia 'Rosemary Webb'. *Below, Matthiola incana* (stocks) well-grown under glass.

Jescot Sonnet – velvety-red, striped with deep maroon
Jescot Valna – pink-mauve with white tip, strong growth
Lady Tweedsmuir – delicate lilac on long stems; good shape, very floriferous
Love's Dream – light raspberry-red shading to pink
Muriel Gladwell – tomato-juice red
Polly Peachum – outer half of florets fuchsia-purple overlaid with tyrian-rose; inner half tangerine-orange overlaid with geranium-lake
Rosemary Webb – almond-blossom pink, paeony-shaped flower
Spring Morning – a rose-pink shading through salmon to gold at centre
Tu-Tu – purest white of good form
Yellowhammer – clear yellow

GERBERA JAMESONI (Transvaal Daisy)

A very showy plant, native of South Africa. The leaves are 6–12 in. long, hairy and lobed. The flowers, which are borne on long slender stems and may be as much as 4 in. in diameter, are daisy-like, and may be any shade of yellow, orange, red, pink or white. There are doubles and singles.

Cultivation. Gerberas are propagated by seeds which should be sown in late January or early February in Levington or Alex seed compost. The seeds, which are rather flask-shaped, are quite large and so may be sown singly, spacing them out in the boxes. There is a slight hollow at the top of the seed, and it is very important to make sure that water does not collect in this after sowing, otherwise the seed may rot.

As soon as they are large enough to handle the seedlings should be pricked out into further boxes, this time using the Levington potting compost No. 1. They must be handled very carefully, great care being taken to plant shallowly so that the crowns are not covered with soil.

In subsequent years the stock may be increased by division, and this of course is the only method possible where it is desired to increase a particular colour, as the plants do not come true from seed. June is perhaps the best month to do this, as the new stock will then have time to become well established before the winter.

Gerberas are usually grown in borders which should be well prepared beforehand, making the soil as similar to the Alexpeat compost as possible, and attention must be paid to the drainage as gerberas will not tolerate stagnant moisture. When planting or potting, as when pricking out, the crowns of the plants must be kept above soil level. Plenty of moisture is required at the roots in the growing season, but it must be kept off the crowns and foliage as much as possible. During the winter much less water should be given and the soil should be kept on the dry side.

Temperature and ventilation. A temperature of from 40°F should be maintained throughout the growing period and as much air as possible given. Ventilation is an extremely important factor, and great care should be taken not to allow the temperature to fall to freezing point or to rise above 60°F. High temperatures encourage soft weak growth.

LATHYRUS (Sweet Pea)

This well-loved plant is used extensively for producing cut flowers from February to early July. After about mid-July all needs can be met by the plants which are grown out-of-doors. The plant is typical of the family Leguminosae, having fibrous roots covered by numerous nitrogen-bacteria nodules. The leaves are pinnate, and the stems have leafy appendages giving a wing-like appearance. The flowers are borne upon long stems with 2–5 blooms per stem.

Cultivation. Plants are raised in a propagating house from seed which is sown ½ in. deep in 2 in. pots (4 per pot) in Alexpeat seed compost. Any small black and hard-coated seeds such as are produced by certain of the darker varieties should be 'chipped' before sowing to hasten germination.

The first sowing should take place during September at a temperature of 50–55°F. Successional sowings are usually made at two-week intervals until late February. A few days before transplanting, the pots should be moved to the final growing house so as to become acclimatised. Adequate light is extremely important.

The ideal soil is a medium loam that can be shallowly dug. Where the drainage is inferior or non-existent it may be necessary to lay tile drains at a depth of 2 feet, one along the back and the other along the front of the border running the full length of the house. The soil will probably require a good dressing of well-rotted dug-in compost. A high lime content is important to maintain good fertility, so apply ground limestone at, say, 6 oz. to the square yard if there is a deficiency. A dressing of fish manure at 3 oz. per square yard is also necessary and this should be pricked into the top 2 in. of the soil, the lime being applied as a top-dressing afterwards.

The soil in the borders should be firmed and a fine tilth prepared for planting. The plants should be set out in rows across the border allowing alleyways between every two rows to facilitate easy training and cutting. The spacing should be as follows: 4 in. between the plants in the rows and 18 in. between the rows, the alleyway being made by missing one row in every three.

Temperature and ventilation. An average temperature of 50°F should be maintained during the growing period, although this may of course rise to 60°F or over when there is sun heat. Ventilation should be given day and night on all possible occasions when favourable weather occurs, as this will help to control bud drop and mildew.

Watering. Sweet peas require a fair amount of water during growth but overwatering tends to cause bud drop during winter. A thorough syringing every five days during this period is usually sufficient.

Supports. As soon as the plants begin to grow, some means of support is necessary, either using 1 in. gauge mesh cord netting stretched alongside each row, or by means of individual strings to each plant as is practised in tomato growing. The strings are tied to wires running parallel to the rows at the top of the house and the bottom ends tied to further wires stretched along the house at ground level. The plants are then twisted round the strings as they grow.

The tendrils should be removed by snipping them off with scissors, as they grow very quickly and tend to deform the foliage and sometimes the blooms. This applies to the winter-flowering varieties which will bloom before side-shoots are formed. The main-crop varieties, however, are often grown as cordons and this of course entails the removal of all side-shoots before they become too big.

Feeding. Overfeeding, especially with nitrogenous fertilisers, should be avoided. Top-dressings should not be given until the plants are more than 12 in. high and the light intensity good. A balanced organic fish fertiliser containing nitrogen, phosphate and potash in the proportions of about 6:7:6 may be applied every four weeks at the rate of 4 oz. to the square yard.

Bud Drop. This is perhaps the chief difficulty with which the grower of glasshouse sweet peas has to contend. It may develop from a variety of causes, including overfeeding (causing sappy growth), over-watering, over-forcing, lack of ventilation, draughts, or sudden changes in atmospheric conditions, and so on. It is essential, especially during the winter and early spring when the weather is cold, wet and dull, to keep the soil on the dry side. If the plants tend to be at all sappy, some potash should be given to harden them up. This may be applied in the form of wood ashes at 6 oz. per square yard just before watering. Plenty of air should be given both during the day and night whenever possible, though of course draughts should be avoided. The ventilators require constant attention in the spring when the weather is changeable.

MYOSOTIS (Forget-me-not)

The forget-me-not is a dwarf fibrous-rooted plant with small blue flowers with yellow centres. There are also pink and white varieties, but these are not popular for market work. It is a favourite in winter and early spring, particularly at Easter-time.

Cultivation. Propagation is by seed which should be sown outdoors in July in drills 9 in. apart in a shady border. As soon as the seedlings are large enough, they should be planted out into nursery beds and grown on until October. They then should be planted 10–12 in. apart each way in the greenhouse.

Temperature. About 50°F is suitable.

Watering and ventilation. Watering must be done very carefully and plenty of air given, as these plants are very susceptible to mildew.

Species and varieties

M. sylvatica is the usual species grown for winter-flowering and Bluebird, with dark blue flowers, is the best variety

CHAPTER 12

Chrysanthemums

GENERALLY chrysanthemums for the greenhouse are grown in pots. This enables the owner of the glasshouse to grow, for example, a crop of tomatoes in the summer, and when these are over the chrysanthemums are lifted in to take their place.

Those who are very keen on chrysanthemums cultivate them in order to provide flowers month after month, planting them out directly into the soil where they are intended to flower. For instance, for flowering in October the planting is done in June, while for flowering in November planting is carried out in July. There are a number of nursery firms which are glad to supply plants for this system of growing.

Propagation. In the normal way, chrysanthemums are raised by striking cuttings. When the old pot plant has finished flowering, the stems are cut down to within 6 in. of soil level. This shortened stem and the roots below are referred to as the 'stool'. Only the best and healthiest plants should be selected as stools. A week or so after the cutting down, the stools may be hand-forked up or knocked out of their pots and most of the soil removed from the roots. The stools are then stood upright in boxes, as close to one another as possible, on a layer of sedge peat 1 in. deep. Another 1 in. layer of this peat is put over the top of the roots and a light watering given. It is advisable to tie a wooden label with the name of the variety to the main stem of each stool.

The boxes are then put in a frame, so as to be protected from frost, and brought into the greenhouse at the end of January or February, and placed on the staging, with the temperature at about 40°F.

The alternative to planting the stools in boxes is to plant them upright, no deeper than half an inch, in the soil of the greenhouse, where they can get plenty of light. A 1 in. layer of sedge peat may be put over the top of them, and a light watering given. The individual varieties should be kept apart from one another, and carefully labelled.

Whether the stools are in their boxes or in beds in the greenhouses, they should be syringed over from time to time with clean water so as to encourage the production of cuttings. Do not, however, make the soil or peat sodden, and do not allow the temperature of the house to rise above about 40°F for the first few weeks. At the end of a month in the greenhouse, the temperature may be allowed to rise to 50°F.

Some gardeners prefer, in the case of pot-grown plants, not to disturb them at all. They merely cut the stems down to within 6 in. of soil level, keeping the pots just as they are in a deep frame. At that time the pots do not want watering for a number of weeks. No stools should be coddled, and they only need protection from serious frost. In the colder parts of the north of England and Scotland a frame containing chrysanthemum stools should be covered with sacking during hard frosts and be brought into the greenhouse probably about the middle of January.

'Sterilising' with warm water. The stools of plants that have been attacked by the minute eelworm the previous season must be placed in warm water at a temperature of exactly 115°F for 5 minutes to kill the eelworms. In the summer attacked plants show brown curled leaves which later look dead. The infection attacks the lower leaves first, and then travels upwards.

Taking cuttings. The actual month for taking cuttings depends on the type of chrysanthemum. For the late-flowering singles and pompons, it is best to take cuttings in late January and February. For decorative dwarf chrysanthemums in pots cuttings may be taken as late as April and early May. For ordinary winter chrysanthemums in pots cuttings are taken in January and early February. Those who go in for growing large exhibition varieties usually take their cuttings in December and early January.

After some time in the greenhouse, growth should develop round the base of the stool, and when these shoots are about 4 in. long, they can be used as cuttings. To encourage their production, the temperature in the greenhouse may be raised, for a period of a week or two, to 60°F, but once plenty of cuttings are formed, the temperature should be reduced to 50°F. If at this time any greenfly or aphides are seen spray with liquid Derris. The best cuttings come from the shoots growing straight up from the roots. Never take cuttings from the growths developing from the stems. Choose for cuttings growths that are moderately thick and look strong and healthy. Having removed the cutting, use a razor-blade or a sharp-bladed knife to make a cut through the stem just below a node (this is the somewhat swollen part on the cutting from which a leaf grows) reducing it to about 2 in. in length.

Dip the base of this cutting into a hormone powder called 'Seradix B'. Fill a seed box 14 × 8 × 2½ in. with Levington or Alexpeat compost and press it down lightly so that its level is a quarter of an inch below the top of the box. In this you should be able to grow 6 rows of cuttings at 4 to a row. Make ½ in. deep holes with a pencil in the level compost, put the cuttings in position and firm them with two fingers of one hand. When the 24 cuttings have been inserted and firmed, give a good watering through the fine rose of a can. Some people start by putting an $\frac{1}{8}$ in. layer of coarse silver sand over the top of the level compost so that when the pencil or thin dibber makes the necessary holes some of the sand is taken down into the bottom. It is on this that the base of the cutting rests, and it does help to ensure quick rooting.

Place the boxes containing the cuttings on the staging of the greenhouse at a temperature of 60°F. If this temperature is not easy to achieve, put a large bottomless box on to the staging first and into this put a 2 in. layer of damp sedge peat. Put the box of cuttings on top of this, then cover the top of the bottomless box with a plain sheet of glass, and this will provide extra warmth.

The cuttings should be rooted by the end of a fortnight. When this occurs, the boxes may be put on a shelf nearer the light for a fortnight, and after that they can go into a frame for hardening off. As a general rule, this takes place about the middle of March. Though chrysanthemums are fairly hardy, they must always be protected from late spring frosts which can damage the growing points of the plants.

Potting up. Once the cutting has produced a nice number of roots, and has been hardened off in a frame, it can be potted up into a 3 in. pot using the Levington or Alexpeat compost No. 1. A crock should be placed over the drainage hole of the pot, concave side downwards, the pot should then be half filled with compost and be tapped lightly on the bench to firm. The well-rooted young plant is then placed on the compost in the centre of the pot and more compost poured around. Further tapping on the bench helps settle the compost evenly. The compost is then pressed down with two fingers until the level is half an inch below the rim of the pot.

The potted up chrysanthemum plants can now be placed in the frame, where they can get the necessary protection from excessive rain and frosts. The plants should be syringed overhead with clean water at about 10 a.m. each day for about ten days, and it is after this period that a good watering will become necessary. By raising the back of the frame on a brick during the fine days the plants will get plenty of air and should grow strong and sturdy. By the beginning of May in the south, and by approximately the middle of May in the colder parts of the north, it should be possible to remove the frame lights altogether.

Those who have no frame but do have a second greenhouse can stand the pots on the staging of this house at a temperature of 40°F. Here the ventilator should be opened on all warm days and, as the temperature gets warmer it may be possible to stop heating the greenhouse altogether.

Potting on. The late-flowering varieties should be potted up into their 3 in. pots in the third or fourth week of February. Never do the first potting too firmly or else there will be difficulties when potting on into the 6 in. pots, which may take place about the third week of March. Where the potting up into the 3 in. pots is not done until the fourth week of February, the potting on will take place at the end of March, but it should always be done before the roots become thoroughly matted in the 3 in. pots. The gardener calls this condition 'pot bound'.

As the potting on is done, look for signs of insect pests, and particularly aphides, and if they are seen dip the plants, pot and all, into a solution of nicotine and soft soap. The formula should be $\frac{1}{4}$ oz. of liquid nicotine and 1 dessertspoonful of soapflakes to a $2\frac{1}{2}$ gallon bucketful of water. Having treated the plant, turn it upside down, putting two fingers on either side of the stem, and tap the rim of the pot on the edge of the bench. The ball of soil should then come cleanly out of the pot.

Remove the crock from the bottom of the ball, and in the new clean 6 in. pot put in a large piece of broken crock, curved downwards over the drainage hole, and half a handful of smaller crocks over the top (not necessary in plastic polypots). Over this put half a handful of rough sedge peat, and on top place the Levington or Alexpeat compost No. 2 so that it half fills the pot. Firm this compost with the fingers, and stand the ball of soil from the 3 in. pot on top of the compost, right in the centre of the pot. Place new compost all round the outside of the ball of soil and, with a 6 in. long stick about the thickness of a broom handle and tapered a little to the base, ram the compost down firmly all round. Add more compost on the top so that there is about half an inch of covering above the top of the ball of soil. The final soil level should now be 1 in. from the rim of the pot.

Put the pots containing the chrysanthemum plants back into the cold frames, standing them as close together as possible, and put the frame lights over the top, as this will help them to recover quickly from the potting on operation. After four days, remove the frame lights and water carefully, making sure the compost never becomes sodden.

Pinching out or stopping. About the end of April, pinch out the growing points of the plants or follow the instructions, if any, given by the nurseryman in his catalogue. Reflexed decoratives, for instance, are often stopped about the middle of April.

The second potting on. It should be possible to pot on the plants from their 6 in. pots to 8 or 9 in. pots during the last two weeks of May. Those who feel they can not afford such large clay pots may use whalehide pots instead. These only last one season.

If using earthenware pots, put one large crock over the drainage hole and then a handful of coarse sedge peat. The pot should then be half filled with Alexpeat compost No. 4 or the Levington compost, and the plant knocked out of the 6 in. pot, placed in the centre with plenty of new compost packed around the ball and pressed down firmly. Turn the pot round and round, adding fresh compost as you do so and ramming all the time.

The top of the ball of soil should be ½ in. below the surface of the new compost when you have finished.

Push two bamboos into the compost on opposite sides of the pot in such a way that they slant slightly outwards (this is because the plant is going to spread at the top). The height of the canes after they have been pushed in should equal the height of the chrysanthemums when fully grown.

Standing the pots out. Find a level spot as near the greenhouse as possible, and cover with coarse broken clinker or black polythene. This prevents the plants rooting through the bottom of the pots into the soil below, as well as ensuring that worms do not work their way up through the drainage holes into the compost.

The site chosen must be naturally sheltered, but if it is not it pays to put up a wind break about 6 ft. high made of hessian and nailed on to strong posts. This need only be a temporary structure and should be sufficiently far away from the plants so as not to shade them.

Put the pots on the ground so that the rows run north to south. Have two rows with the pots touching one another and allow 3 ft. in between these double rows, so that the gardener can walk up the paths to do the necessary watering, feeding and spraying. Drive posts into the ground at the ends of each double row and exactly in the middle, so that they are protruding from the soil about 4 ft. 6 in. and nail a 2 × 2 in. wooden batten 15 in. long to the top of the post so as to form a T-piece. Attach wires to the outside edges of the battens and stretch tightly along the tops of the rows. The tops of the bamboos are then tied to the wires to prevent them from being blown over.

General care. Water about once a week, giving a good soaking. As the weather gets sunnier it may be necessary to water every few days, and in an extraordinarily dry August water at least once a day. Automatic watering devices such as the Tricklematic or Hozelock schemes do save time. The moment the bulk of the roots have penetrated into the new compost, feed with Marinure, starting about 20th June and feeding every ten days up to flowering time.

Keep the branches upright by tying them loosely to the canes. The best method is to have a double twist round the bamboo itself with a loose loop tied around the stem. If the tops of the pots get weedy, hand-weed them and put the weeds on the compost heap with a sprinkling of fish manure.

Some varieties will need to be pinched or stopped a second time, probably about the second or third week of June, and by that time 4 good side-growths should have developed. It is after these are pinched back that the 10 or more good branches will grow, bearing the flowers.

Lifting-in time. Those who live in the northern, colder, frostier parts will have to bring the chrysanthemums into the greenhouse about the third week of September. As a rule southerners need not lift in until sometime in October. As you carry them into the greenhouse, remove the yellowing or dead bottom leaves and spray the plants thoroughly with Karathane to prevent mildew, and liquid Derris to kill any lurking insect pests. The pot plants should be stood on the floor of the greenhouse, or if there is the necessary head-room, on the staging, allowing about 12 in. of breathing space on all sides. The moment the plants are in the greenhouse, give them a good watering.

The temperature in the greenhouse should now be 50°F, and whatever the heating system used, it must keep the air in the house circulating. See that the ventilators are kept open a little in the daytime unless it is frosty or foggy. This will prevent mugginess in the greenhouse and the humidity that would encourage disease. Once the pot chrysanthemums are housed, the pests and diseases may be controlled by using one of the little canisters known as 'smokes', which produce the particular vapour necessary to destroy the infection.

Varieties. There are a very large number of excellent varieties of Greenhouse Decoratives to grow in pots. Unfortunately today's varieties soon go out of fashion, and I have therefore listed a few easy-to-grow 'old stagers' that should give great satisfaction to the beginner.

NAME	COLOUR	REMARKS	BLOOMING
Christmas Carol	crimson, scarlet	reflexed	December
Christmas Wine	wine purple	reflexed	December
Elizabeth Burton	shell pink	incurving	December
Fred Shoesmith	white	incurving	Nov/Dec.
Golden Favourite	rich yellow	reflexed petals	Nov/Dec.
Golden Sussex	bright bronze	reflexed petals	Nov/Dec.

Mayford Perfection	warm salmon	incurved petals	Nov/Dec.
Rose Mayford Perfection	bright rose	incurved petals	Nov/Dec.
Shirley Late Red	Indian Red	reflexed petals	Nov/Dec.
Yellow Mayford Perfection	bright yellow	incurved petals	December

Dwarf Pot Chrysanthemums

For people with small greenhouses, there is a lot to be said for growing plants with stems about 12 in. long. They are easy to handle, not easily blown over on the standing ground, and it is often possible to produce twelve good blooms per pot.

Propagation and potting. Cuttings should be taken during late March and early April as described on p. 134. When the cuttings have rooted, they must be potted up individually into 3 in. pots using the Alexpeat compost No. 1 or Levington compost. I have taken cuttings as late as early May, and potted them up when rooted in June.

From the 3 in. pots the plants are potted on into 5 in. pots at either one per pot or, for a more brilliant show, three plants per pot, using a potting compost No. 3. Pinch back the growing points of the plants on or about 15th May, and carry out a second pinching back or stopping about 15th June. Postpone these two stoppings by about a fortnight in cases where the cuttings are struck later. It is a good practice to stop severely, i.e. to cut back an inch of each growth with a sharp knife. Four laterals or side-growths should develop as the result of the first stopping, and possibly four more as the result of the second stopping, thus resulting in 16 good blooms per plant. This should be the maximum.

General Care. Look after the plants in a similar manner to the taller varieties in their 9 in. pots. They should be spaced 15 in. apart on the standing ground because they bush out so much as they grow. Use one bamboo, pushed into the centre of the pot, per plant, and loop the stems loosely making a tight tie on the cane. There will be no need for posts and wire supports.

Curiously enough these dwarf plants seem to be particularly prone to pests and diseases, and regular spraying with Macuprax may be necessary once a fortnight from June onwards. Spraying with Malathion, which is a systemic insecticide, ensures immunity from pest attack for about three weeks.

Feed with Marinure, but use a high potash type, like Tomato

Special, and the leaves will tend to be firmer. For feeding frequency follow the general instructions on p. 217.

Varieties. These dwarf flowering chrysanthemums must not be confused with the pot chrysanthemums sold in the shops and produced artificially by keeping the plants in the dark for certain periods.

NAME	COLOUR	REMARKS	BLOOMING
Blanche Poitevine	pure white	3 ft. incurved	Nov/Dec.
Yellow Morin	bright yellow	3 ft. free-flowering	Nov/Dec.
Marie Morin	white	3 ft. semi-incurving	December
Pink Morin	shell pink	3 ft. incurving	November
Touchstone	primrose	3 ft. incurving	November
Princess Anne	pale pink	3 ft. decorative	December
Spanish Lady	claret-red	2 ft. anemone centred	Nov/Dec.

Charm Chrysanthemums

This is another group of dwarf chrysanthemums which can be grown in pots and flower from September to December. The plants do not have to be stopped or disbudded, and they produce a mass of flowers which should last for three or four weeks.

Propagation and potting. It is possible to buy seed which can be sown in the Alexpeat or Levington compost during January or February. When the plants are 1 in. high, pot them up individually into 3 in. pots, and about 6 weeks later into 5 in. pots. At the end of May put them outside on the standing ground where they should grow until the end of September or early October. Some people prefer to grow the plants in 8 in. pots – 2 per pot – to make a better show.

The alternative to seed sowing is to buy rooted cuttings, which are treated in a similar manner to the normal dwarf chrysanthemums.

Varieties. With seed sowing one gets a mixture of colours in these open star-shaped flowers, but with cuttings, of course, it is possible to buy different named kinds.

NAME	COLOUR	REMARKS	FLOWERING TIME
Apricot Charm	apricot	3 ft. dainty flowers	Nov/Dec.
Bullfinch	crimson	2 ft.	Nov/Dec.
Flamenco Girl	crimson	2 ft. anemone centred, free-flowering	December
Kingfisher	carmine	2 ft.	Nov/Dec.
Morning Star	lime yellow	2 ft.	Nov/Dec.
Seagull	white	2 ft.	Nov/Dec.
Tang	tangerine	2 ft.	Nov/Dec.

Cushion Koreans

Most Korean chrysanthemums grow happily out-of-doors and flower in August and September. The Cushion Koreans start flowering the second or third week in October, and continue until late November.

Propagation and potting. Cuttings may be taken about 5th March and dibbled out about 1 in. apart into an Alexpeat compost in a frame. Cover the frame with its light, water from time to time, and the cuttings should produce roots by about 5th May, when they may be potted up into 3 in. pots. Some prefer to leave the plants in the frames until a fortnight later, and then pot them up directly into 5 in. pots.

General care. After potting up, the pots can be plunged in soil in the garden 15 in. apart each way and left there to grow until the end of September, when they should be removed. Roots that have penetrated through the drainage hole at the bottom of the pots should be cut off and the pots wiped with a damp rag before being brought into the greenhouse. Once the plants are in the greenhouse, feed them with diluted Marinure once a week until the flowers open. The plants will flower longer if dying blooms are cut off with scissors every 10 days.

After flowering, stand the pots out-of-doors in the shelter of a wall or fence, cutting the stems back to within 1 in. of soil level. Water the pots from time to time, for these are the plants which are to produce cuttings in the following spring. Each plant should produce at least five cuttings.

Varieties. I have had success with the following varieties, but new kinds do become available from time to time.

NAME	COLOUR	HEIGHT	BLOOMING
Bonfire	double scarlet-red	18 in.	October
Glow	semi-double salmon-rose	15 in.	Oct/Nov.
Startler	bright claret-pink	15 in.	October
October Charm	bright red single	18 in.	mid-Oct/Nov.

Single Flowered Chrysanthemums

Most people like large double chrysanthemums, but some prefer singles with a large central eye.

Propagation and general care. Cuttings of the single November/December flowering varieties are taken in exactly the same way as the double chrysanthemums (see p. 134). The general run of November/December singles do not have to be stopped, but if on the other hand the plants haven't started to produce side-shoots by about 18th May, the growing points of the plants should be pinched out. There are a few varieties that need stopping at the end of April, but the nurseryman who supplies them should give advice in this respect.

Those who want to have large single blooms should carry out disbudding, which entails regularly pinching out with the thumb and forefinger the tiny flower-buds growing at the sides of the main terminal ones, thereby producing five to ten good blooms per plant. On the whole, singles are easier to grow than doubles, but the general instructions given on pp. 133–8 apply equally well to both plants.

Varieties. The flowering period of good late singles is November and December. Most varieties disbud well, but those listed as spray types I have discovered are not good disbudders and should be left to grow naturally.

NAME	COLOUR	REMARKS	BLOOMING
Albert Cooper	large yellow	4 ft. fine flowers	Nov/Dec.
Alliance	terracotta	4 ft.	November
Crimson Crown	crimson	4 ft.	November
Desert Song	terracotta with gold tips	3½ ft.	Nov/Dec.
Edwin Painter	apricot yellow	3 ft.	November
Lilian Jackson	rose-pink with white centre	3 ft.	
Midlander	salmon bronze	4 ft.	Nov/Dec.
Phyllis Cooper	rich golden-yellow	4½ ft.	Nov/Dec.
Peggy Stephens	large golden-yellow	4 ft.	Nov/Dec.
Red Desert Song	crimson-scarlet	3½ ft. large flowers	Nov/Dec.
Sylphide	piuk	4 ft.	November
Venetian Glass	shining golden-amber with orange trace	4 ft. petals slightly fluted	Nov/Dec.
Woolman's Glory	terracotta	4½ ft. fine flowers	Nov/Dec.

CHAPTER 13

Carnations

THE commercial carnation grower grows his plants at soil level in large glasshouses. The amateur of course finds it better to grow the plants in pots, and uses for this purpose either the Alexpeat or Levington compost. It is always best to start the plants off in a compost No. 2 rather than No. 1, as this enables you to reduce the number of pottings-on to an absolute minimum. It pays every time to plant the rooted cuttings straight into the pots in which they are to grow for the rest of their life.

Propagation. It is essential that carnation cuttings have sufficient food reserves stored in them to enable them to remain alive while new roots are being produced, and small shoots should therefore be selected from the sides of fully grown plants any time from the middle of October to the end of February. Always choose completely healthy plants from which to take this propagating material. Never take cuttings from a plant which is suffering from drought. Such cuttings root very slowly. Also never take them from the tops of plants or right from their base. The best cuttings come from about half-way down.

The normal cuttings should be about 4 or 5 in. long, and they should be severed from the side of a mature plant with a sharp knife. A cut should then be made with a razor blade at a node (the swollen part on the stem from which the leaves grow) and the curly leaves on the bottom inch of the cutting trimmed off carefully, so that no foliage will touch the sand when the cutting is pushed into the compost.

Having removed, say, three pairs of lower leaves, push the cutting into the sharp silver sand and medium-grade sedge peat, so that its base is about ¾ in. deep in this open compost. Alternatively, you can use sharp silver sand to which fine sedge peat is added, so that the latter makes up a fifth of its bulk. By the way, it is important never to use the same sand a second time for cuttings because, if you do, many of them will fail to root.

The rooting material in which the cuttings are struck should be at least 2 in. deep with really good drainage below. Put plenty of well-washed old crocks underneath the silver sand and peat and if possible have some bottom heat, using the electric wire method described on p. 35.

Generally speaking the cuttings will root in 28 days; I have been lucky in getting cuttings to root in 23 days, but some varieties are obstinate and take 35 days. I find the cuttings strike well in a propagator on the greenhouse staging. It is quite easy to make a hole with a pencil or thin dibber ¾ in. deep and to put the cutting in this hole so that it rests firmly on the base. Firm with two fingers round the cutting afterwards. Carnations can be set in the propagating frame as close as 1½ in. apart. After they are in position, give a light watering, but after this give just enough water to keep the propagating material moist. Those who have not done any propagating before may like to dip the base of the cuttings in a hormone powder such as Seradix to encourage quicker rooting.

Potting up. Once the cuttings have struck well, they may be potted up into 3 in. pots, using the Alexpeat potting compost or Levington compost No. 1. A good crocking is necessary if clay pots are used. On top of this place a handful or two of compost, and then plant the cutting in the centre of the pot so that when covered with further compost it will be at about the same depth as it was in the propagating bed. Do not over-firm, just tap the pot on the potting bench to help settle the compost, and then press down with two fingers of one hand. After giving the plants a normal watering, put them on the greenhouse staging at a temperature of 60°F. Grow them on here quietly for about a month, and then pot them on into 6 in. pots, this time using the Alexpeat or Levington compost No. 2. Once again be sure to put plenty of crocks in the bottom of the pot for good drainage, and cover these with half a handful of really rough sedge peat. Over the top put in two good handfuls of the potting compost No. 2 and press down lightly. Stand the ball of soil in the centre of the 6 in. pot, so that the upper part is about ½ in. below the rim, and fill up with compost. Keep turning the pot as you do so, so that the compost is poured in evenly all round. Firm the compost with a potting stick so that it is at least as firm as the ball of soil, otherwise the water will trickle round the outside and fail to thoroughly moisten the central ball itself.

Those who like to keep their perpetual flowering carnations growing for several years will pot the plants on again at the end of 12 months into 8 in. pots, using this time Alexpeat or Levington compost No. 3. I must point out that the best blooms are always obtained in the first year, but quite good blooms will be produced the second season.

Stopping. It is important to pinch out the growing points of the plants in order to encourage them to produce sturdy branches from their base, and it should be done directly the plants have started to grow in

1. *Cuphea ignea*, p. 102
2. *Cypripedium* 'Clair-de-Lune', p. 149
3. *Dahlia* 'Gerry Hoek', p. 128
4. *Eucomis comosa striata*, p. 151
5. *Erythrina crista-galli*, p. 104
6. *Exacum affine*, p. 83

their 3 in. pots, leaving only 6 pairs of leaves on each plant. After the plants are established in their 6 in. pots, the secondary growths must be stopped when their stems are about 6 in. long. This involves removing the tip together with one pair of properly formed leaves. Some people say that you should not stop all the laterals on one plant at the same time, but should spread the process over several days.

If, however, you want plants to bloom the first season, no stopping should be done after the beginning of August. Those who find that the greenhouse is rather full in the summer can, if they wish, put the pots of carnations into a cold frame on a layer of coarse ashes. The ashes are there not only to provide drainage, but to prevent worms from climbing up through the drainage holes. In the sheltered frame the plants must be syringed over twice a week to discourage attacks of red spider.

Every effort must be made to prevent the stems of the carnations flopping about. To get the best flowers, the stems must be sturdy and upright. It pays to provide support in the form of short sticks pushed into the compost or to use one of those special metal rod and wire supports which some of the carnation nurseries sell for the purpose.

Those who aim to keep their plants a second year should realise that the season starts about the middle of February, even though the potting on is done at the end of June or early July. Weed care, trimming and fortnightly feeding with Marinure may be done from the end of February onwards, and where the plants are in 8 in. pots, it is usual to give an additional teaspoonful of carbonate of lime per plant. In about the middle of October in the third year the plants are scrapped and the gardener starts all over again. Many people don't even bother to keep the plants as long as this.

Ventilation. Carnations like plenty of fresh air, and the ventilators should be kept open at night-time in the summer when temperatures are high. Because it is necessary to start with a moderate amount of ventilation at, say, 9 a.m., to increase it at mid-day, then to close a little if the sun goes in, and so on, many people prefer to use automatic ventilators (p. 39).

Shading. Keep the glass absolutely clean where carnations are growing. Space the pots out at least 6 in. apart, so that the plants have room to breathe. Don't shade the house in the spring or early summer for carnations love sunshine, especially if you give them plenty of ventilation. I have used a little shading at times when the sun was very hot and bright in September.

Do not coddle carnations even in the winter, but keep the temperature at about 45°F. When carnations are grown in plastic polypots they only need about a quarter the amount of water, because the sides of the pots do not lose moisture like clay ones. Naturally, more water is given in the late spring and summer than in December, January and February.

Varieties

PINKS
Dusty Sim – dusty pink
Flamingo Sim – outstanding pinky salmon
Laddie Sim – a strong salmon shade
SCARLET
William Sim – free-flowering, red
HELIOTROPE
Richard Dimbleby – good form, of a deep claret
YELLOWS
Harvest Moon – a sport of William Sim, yellowy-orange
Tangerine Sim – orangey-yellow
FANCIES
Cocomo Sim – yellow, heavily striped crimson
F. G. Garnett – free-flowering, yellow striped pink, large
Zuni – vigorous growth, deep cerise flecked maroon
WHITES
White Sim – free-flowering
CRIMSONS
Joker – erect and wiry
Topsy – free-flowering, deep crimson

CHAPTER 14

Simple Orchid Growing

ORCHIDS are extremely popular flowers but growing them is regarded as a specialised job and is carried out by only a small number of amateurs. Orchid growers may be divided into two groups – amateur growers, and growers who supply plants retail and sell the flowers when available.

The following are the most useful types for cut blooms:

CATTLEYAS. These form two to three large flowers on a shoot, mauve being the predominant colour. The flowering period is, however, not so long as that of cymbidiums.

CYMBIDIUMS. These form long sprays bearing a number of handsomely coloured flowers.

CYPRIPEDIUMS. The members of this species are often used as buttonholes, especially in winter when they are in abundant supply. The main flowering period is from November to January. The principal ones grown are:

- C. *insigne* (the original wild form), which has greenish-yellow flowers spotted with chocolate
- C. *mauliae* a fascinating unusual browny colour
- C. *sanderae* (lime-yellow), and various hybrids

ODONTOGLOSSUMS. These are somewhat similar to cymbidiums in their spray form. The colour range is very varied, the flowers being generally splashed and spotted with red. The pure white *O. crispum* is very popular.

Propagation. The usual method of obtaining new plants, of course, is to buy them, but when plants have developed a double 'lead' or shoot, it is possible to propagate by division. This should be done after the resting period.

Propagation by 'seed' is extremely difficult and is usually only carried out by specialists, because the seeds of the orchid will only germinate in conjunction with the spores of certain fungi. Usually the seed is sown, together with a small quantity of the fungus, in seed pans of sterilised soil and leaf-mould mixed in equal parts and kept in a greenhouse at a temperature of 65–70°F. When saving home-grown seed it is necessary to allow the seed pods to remain on the plant for at least nine months to allow the seed to ripen fully. After about twelve months the plants may be potted up into 2 in. pots, using a compost consisting of moss and osmunda fibre in equal parts. They will then flower in about seven years – a slow job!

By far the easiest method of propagation, therefore, is by division.

All new plants or divided specimens should be thoroughly cleansed with petroleum emulsion to destroy 'scale' insects, and all dead pieces should be removed. They are then usually potted up into 6 in. pots containing a suitable compost for the species, and, as they grow, they may be moved into larger pots.

Temperature. Cattleyas require a temperature of 60–65°F; cymbidiums require a minimum of 45–50°F; odontoglossums need a temperature of 50–55°F; cypripediums will grow quite well at a temperature of 50°F, while for modern hybrids 60°F, and for mottled leaf types 65°F are the optimum temperatures.

Ventilation. Adequate ventilation is necessary at all times, taking care to maintain the correct temperature.

Shading. Shading is also necessary to prevent scorching during bright sunny weather – many gardeners use indoor blinds for the purpose.

Cultivation and potting composts. Any sort of house is suitable for the cultivation of orchids, preference being given to the low 'pot-plant' type house with the ridge about 8 ft. high. It is usual to have a central path with wooden staging on either side, under which are side ventilators.

When potting on into the final pots, the different species vary somewhat in their compost requirements:

CATTLEYAS. Because of the high temperature they require, and the cost of fuel, these are no longer the most popular orchids in Britain, although they do produce large, handsome and most colourful blooms. Compost in pots: 1 part osmunda fibre, 1 part sphagnum moss, a sprinkling of crushed oak leaves.

CYMBIDIUMS. These are now the most popular orchids in this country, due to their low temperature requirement and the durability of the plants and flowers.

Keep the compost moist throughout the year, using rainwater if possible. Keep the atmosphere in the house moist by damping down paths and under the staging twice a day in summer and once a day during winter.

Cymbidiums can be grown in pots or boxes. Those in large boxes (say 15 × 15 × 6 in.) are more prolific, giving two or three growths from each leading bulb, increased flower production by one-third per spike and many more spikes per plant.

Through the new method of Meristem propagation, the finest and most modern hybrids can be obtained at a much lower price than ever before, and the amateur grower can purchase plants produced by this specialised method from The Stonehurst Nurseries, near Ardingly, Sussex.

CYPRIPEDIUMS. 1 part osmunda fibre, ½ part loam fibre, 1 part sphagnum moss.

ODONTOGLOSSUMS. As for cattleyas, ½ part silver sand and plus another ½ part sphagnum moss, ½ part of clean dry beech or oak leaves.

Final potting. The plants in the large 3 in. or 6 in. pots should be potted on just as root growth commences; cypripediums in late winter after flowering, and the others during spring. Further pottings or top-dressings with a fresh mixture may be carried out every two to three years.

The actual potting consists of placing crocks to a depth of about one-third of the pot, covering these with compost and firmly potting the plant. The pots are then set out upon the staging, allowing sufficient room between the plants to prevent the leaves touching. Routine watering is necessary and the plants should be sprayed over during the growing season.

After flowering, orchids should be partially dried off for a few months before starting them into growth again, the exception being cypripediums, which have no pseudo-bulbs to retain the moisture.

Insecticides. Use Kelthane for controlling red spider and Malathion for greenflies and thrips.

CHAPTER 15

Bulbs and Corms

THERE are large numbers of different types of bulbs and corms that can be grown in the greenhouse. With most of them you should pot the bulbs up in fibre or sedge peat, keep the bowls or pots in the dark for nine weeks or so, and then bring them out into the light, for instance on the staging of the greenhouse. The reason why bulbs are kept in the dark for nine or ten weeks is that it is tremendously important to get them to produce a large root system *before* attempting to produce leaves and flowers.

Into the category of this method of culture – often called 'forcing' because the bulbs can be made to flower earlier – can be placed daffodils, tulips, hyacinths, crocuses, and so on. There are also bulbs and corms that are grown in Levington compost or Alexpeat compost year in, year out, and are not forced in the true sense of the word – they are grown in the greenhouse because they do not grow successfully out-of-doors. Into this category go amaryllis, cyclamen, *crinum* (Cape lily) and *eucharis* (Amazon lily).

Forcing Culture

BABIANA South African bulbs rather like ixias.

Cultivation. Crock a 6 in. pot well and fill with medium-grade sedge peat. Place six small bulbs in the peat so that they are just covered. Grow at a temperature of 50°F and keep them in the dark for 8 weeks before bringing into the light. Bulbs potted in autumn flower in April.

Species. Blooms are red and orange. The best species is perhaps *Babiana rubro coerulea* – a red and blue as its name suggests.

CHIONODOXA (Glory of the Snow). These plants, native of Crete and W. Turkey, only grow 8 in. high and flower early bearing blue, white or pink flowers.

Cultivation. Buy the bulbs as soon as they become available in the autumn. Pot them up in bowls or pots $1\frac{1}{2}$ in. apart and so that the tips of the bulbs are showing above the soil. Put them in the dark under sand, peat or ashes for at least nine weeks.

Species and varieties

C. luciliae – cambridge blue
C. luciliae Pink Giant – tyrian-rose
C. luciliae rosea – soft parma-pink

CROCUS The crocus is extremely useful as a pot plant for early spring, and it is very easy to get it in flower at the same time as the early forced pots of daffodils and hyacinths. Crocuses are obtainable in a variety of colours – yellow, purple, blue, white and striped.

Cultivation. The corms should be bought in August or September from a reputable dealer. They should then be planted in 3 in. pots, two or three per pot, great care being taken to set them with the pointed end uppermost. They should be buried to about half their depth in a light open compost, such as Levington compost. The pots should then be plunged in sand or ashes, with the tops 5 in. below the surface. Water with a solution of Cheshunt Compound to prevent rotting and disease attack just before the soil is placed over them.

As soon as growth commences in spring, the pots should be lifted into the greenhouse, and maintained at a temperature of 55–60°F.

Watering should be done with discretion, picking out only those pots or bowls which have become dry.

Varieties

Kathleen Parlow – pure white
Little Dorrit – silvery-lilac with purple blotch
Remembrance – deep doge-purple
The Bishop – a deep rich purple
Yellow Giant – golden-yellow

EUCOMIS An unusual African bulb bearing thick spikes of greenish and brownish flowers.

Cultivation. Plant three bulbs close together in medium-grade sedge peat in a well-crocked 6 in. plastic pot. Likes a temperature of 50°F. When flower-spike first shows, give Marinure once a week. Never force. Pot in October for flowering in April.

Species

Eucomis comosa, the pineapple flower – flower spike 1 ft., flowers yellowish-green, scented

FREESIA The freesia is an extremely useful pot plant during the months of December and January, and often lasting until March. The white species *refracta*, variety *alba* is extremely pretty, but has been superseded by the delicate colouring of the Bartley strain.

Cultivation. Freesias may be propagated both by means of bulbs and seed.

BULBS. Bulbs should be planted in 6 in. pots during early August, spacing them 1 in. apart and allowing the crowns to appear just above the soil. The compost generally used consists of 4 parts good loam, 1 part lime rubble, 1 part well-rotted manure or compost and 1 part coarse sand. Alternatively, the Alexpeat potting compost No. 2 may be used.

When the potting is completed the pots should be stood in a sunny cold frame but *no water* should be given until growth starts. In fact, should any rain occur, the frame lights should be put on to the frames immediately. If water is given at this stage only spindly growth and premature flowering results.

About the end of October, the pots should be transferred to a greenhouse and placed upon the staging in a steady temperature of 55°F. From this time onwards routine watering should be carried out, care being taken never to allow the pots to dry out. Light twiggy stakes may be necessary to support the foliage.

Thrips sometimes prove very troublesome, but nicotine fumigation will soon control this pest.

SEED. Freesias raised from seed, especially the coloured species, are becoming quite popular among gardeners. The seed should be sown in March in seed trays filled with the Alexpeat or Levington seed compost, and a temperature of 60°F should be maintained during the first few weeks.

As soon as the seedlings are big enough to handle, they should be pricked out into further boxes, one inch apart, using the potting compost No. 1. Later, before they become too big, they should be pricked off into 6 in. pots, again 1 in. apart. During early June they should be ready for transferring to cold frames. From this time onwards the treatment should be as described for the bulbs (see above).

Varieties

Blue Banner – lobelia-blue with white throat
Orange Favourite – saffron-yellow with orange glow
Sapphire – a delicate shade of blue, merging down to white
Souvenir – a beautiful creamy gold

White Swan – a large pure white
Stockholm – carmine-red with gold throat

GALANTHUS (Snowdrop, Fair Maid of February)

Cultivation. Pot up in October or November in sedge peat in pots or bowls. Plant bulbs 2 in. deep and 1 in. apart, and 6–8 bulbs can go into a 6 in. pot or bowl. Place pots out-of-doors covered with sand or peat. Remove after eight or nine weeks and bring into the greenhouse on the staging. Water only moderately.

Species and varieties

G. elwesii – a beautiful species, large globular snow-white flowers, inner segments marked with rich emerald-green
G. nivalis – the Common Snowdrop
G. nivalis flore pleno – a double form of the common snowdrop
G. nivalis maximus – large-flowering, vigorous
G. nivalis S. Arnott – Giant Single Snowdrop, beautiful hybrid with large snow-white globular flowers, inner segments marked with rich emerald-green, tall, scented
G. nivalis viridi-apice – large globular pure white flowers, inner segments and petal-tips marked with rich emerald-green, vigorous grower

GLADIOLI The small hybrid species *colvillei* has become a very popular pot plant because of its similarity in shape and colour to the larger species.

Cultivation. These small-flowered plants are grown from dry corms in much the same way as freesias. The corms should be potted up into 6 in. pots during August, burying them about 2 in. deep. If the corms are small, two may be used per pot. A similar compost to that used for freesias is suitable, i.e. 4 parts good loam, 1 part lime rubble, 1 part well-rotted manure or compost and 1 part sand. Alternatively, potting compost No. 2 may be used.

The pots should be buried 6 in. deep in beds of ashes, peat or sand, and left until early in January. They should then be removed in succession from the frames and moved on to the greenhouse staging at an even temperature of 55°F. Water should be given when necessary during this forcing period so that the pots remain reasonably moist.

If any of the plants become too tall, a light stake such as a pea stick should be given for support.

It must be remembered that light is extremely important, so forcing should not be practised too early in the year when the light intensity is low, as the plants tend to become blind and fail to flower.

DWARF GLADIOLI These are very suitable for pot culture and wonderful for table decoration. They last for a very long period in full bloom and often throw up second spikes of flowers when the first one is nearly over. These spikes usually grow to a height of 18 in. and sometimes to 2 ft.

Cultivation. The potting up of dwarf gladioli can be done in the late autumn, but there must be no attempt to bring the pots into heat in the greenhouse until March. The pots can on the other hand be brought into a frame or cold house late in January.

Species and varieties

Blushing Bride – pure white, flecked with pink and carmine
Charm – a delicate lilac
Rose Precoce – an attractive salmon
Spitfire – scarlet-red with purple blotches
Suzanne – a pearly pink with cream splash on lower petals

HYACINTH As a pot plant, the hyacinth is extremely popular from Christmas until March. The plant is typical of its natural order in having a scaly bulb with a papery skin. The leaves are lanceolate and waxy with fleshy stems. Flowers are again typical, having joined petals forming a ring. Colours include white, pink, crimson, mauve and blue.

Cultivation. During late August or early September first-grade bulbs may be potted up into large 3 in. pots, or three or four to a bowl, filling the pots to within $\frac{1}{4}$–$\frac{1}{2}$ in. of the top with a compost consisting of good turfy loam together with a little sand to assist drainage (the amount of sand varying, of course, according to the type of loam used). The bulbs should be placed in the soil in such a way that only the bottom half is buried. They should be well watered in with a medium-strength solution of Cheshunt compound in order to prevent rotting due to disease spores in the soil. The pots must then be buried 9 in. deep in sand or peat, and left alone for a few months, so as to let normal growth proceed slowly and a good root system form.

During early December, some of the bulbs which were potted first should be inspected, and as soon as an inch or two of growth has appeared they should be lifted and placed upon the greenhouse staging. These plants should not be forced, but sufficient heat should be maintained at night to prevent 'frosting'. Ventilation may be given on all favourable occasions. No watering will be necessary, and all that remains is to wait for normal flowering.

Varieties

City of Haarlem – deep yellow
L'Innocence – white
King of the Blues – rich indigo-blue
Queen of the Blues – a lovely pastel blue
John Bos – rich rose-red
Princess Irene – ice-pink flushed with rose
Lord Balfour – parma-violet
Winston Churchill – porcelain-blue inside, light Prussian blue outside
Yellow Hammer – creamy yellow

IXIA (African Corn Lily)

Cultivation. Grow in a similar manner to freesias (p. 152). Pot up in September in Levington compost and keep just moist. Do not put in the dark as with other bulbs, but grow on the staging in full light at a temperature of 50°F. Loves plenty of sun when the buds first appear.

Species

Ixia campanulata – purple and cinnamon flowers
Ixia patens – pink flowers

MUSCARI (Grape Hyacinth)

Cultivation. These bulbs may be treated in a similar manner to chionodoxa. They need to go into the dark for nine or ten weeks before being put on the staging of the greenhouse at a temperature of 60°F. They flower early about three weeks after being put into the greenhouse, and the flowering period can be extended by bringing bowls into the greenhouse at intervals.

Varieties

Azureum – soft powder-blue
Blue Pearl – cobalt-blue
Heavenly Blue – bright sky-blue
M. tuberginianum – clear blue

NARCISSUS or DAFFODIL Daffodils are easy to grow as pot plants and have the added advantage that after flowering they may be planted out-of-doors where they will give pleasure for many years to come. The plant is a typical member of the family Amaryllidaceae, having long waxy green lanceolate leaves which arise from a scaly bulb covered by a papery skin. The flowers are borne upon long fleshy stems. The colours are basically yellow and white, although some varieties have red and orange colouring on the trumpet.

Cultivation. Those bulbs that have been graded as 'rounds' or single-nosed are best. During early September the bulbs should be planted, four per 6 in. pot, in a loamy compost with a little sand to assist drainage. They should be only half buried in the soil and fairly firm.

The pots should be placed as close as possible in a shallow trench outside, six inches deep and any convenient length and width. They should then be well watered in with a solution of Cheshunt Compound to prevent the bulbs rotting due to disease spores in the soil. When this operation is complete the pots should be covered to a depth of 9 in. with soil or sand. To assist in removing the soil when the boxes or pots are eventually taken into the houses, it is best to cover with half an inch of peat prior to putting on the soil or sand covering.

The bulbs are now left alone for a few months, so as to let growth proceed slowly while a good root system is being formed. During December some of the earlier batches should be inspected, and as soon as the first inch or two of growth is showing, and its roots are showing through the drainage hole, the pots may be brought into a cool greenhouse and placed upon the staging. Any soil clinging to the leaves should be washed off.

The plants should not be forced but sufficient heat should be maintained to keep out any frost. Ventilation should be given upon all favourable occasions. Only a little watering will be necessary, but the pots will need staking with twiggy sticks as soon as the leaves and flower-stems become heavy, in order to prevent them flopping.

Species and Varieties

YELLOW TRUMPETS
Golden Harvest – rich yellow trumpet and perianth
King Alfred – a rich, glorious yellow
Van Waveren's Giant – golden-yellow

BI-COLOUR TRUMPETS
Forsight – milk white with yellow trumpet
Trousseau – pure white, creamy yellow trumpet

LARGE CUPPED
Lord Kitchener – large white perianth, pale citron cup
Helios – deep golden-yellow perianth, orange trumpet-shaped cup
Fortune – deep yellow perianth, orange-red cup

SMALL CUPPED
Barret Browning – white with flame crown
La Riente – creamy white with brilliant orange cup

POETAZ HYBRIDS

Cheerfulness – three to four large blooms. Creamy-white perianth; double centre of creamy-white and yellow. Fragrant

Primrose Beauty – primrose yellow

TAZETTA AND TAZETTA HYBRIDS

Soleil d'Or – orange cup, golden-yellow perianth. A very popular variety

Glorious – a beautiful flower, pure white with a brilliant red cup

Laurens Koster – pure white, lovely orange cup

POETICUS

Ornatus – double white

Actaea – beautiful, pure white perianth and yellow cup edged with fiery red

Queen of Narcissi – broad snow white perianth, citron cup with deep red rim

DOUBLE

Double Event – pure white rounded edges, centre ruffled yellow

Texas – creamy-gold and tangerine

ORNITHOGALUM *Ornithogalum umbellatum* is the Star of Bethlehem, called in the Bible a 'cup of dove's dung' because of its white flowers.

Cultivation. Does best at a temperature of 40°F. Pot up in September or early October, four bulbs per 6 in. pot. Stand the pots in a cold frame until mid-January, then bring into the greenhouse and place on the staging. Grow on quietly and the blooms will soon appear.

Species. I recommend *Ornithogalum umbellatum*, which grows 1 ft. high. Also *O. pyrenaicum*, which produces yellowy-green flowers later than *umbellatum.*

POLIANTHES Another relation of the amaryllis. A tuberous-rooted specimen with beautiful scented blooms.

Cultivation. Pot up African varieties in October or November to flower in the following autumn. Pot up American varieties in February or March, to flower in winter or spring. Use the Levington or Alexpeat compost No. 2. Firmly plant two bulbs per 6 in. pot, covering them to two-thirds of their depth. Place in a frame and cover completely with peat; do not water. Examine after ten to twelve weeks, and if growth has occurred bring into greenhouse at a temperature of 60°F, reducing to 50°F when in bloom. When growing, syringe the leaves every day and feed once a week with Marinure. The bulbs that arrive in the early spring can be potted firmly in Levington compost and put straight into the greenhouse at a temperature of 65°F. Water well, and when growth begins reduce temperature to 55°F and give plenty of light.

Species. Polianthes tuberosa commonly called the Tuberose, has lovely white fragrant blooms on stems 3 ft. tall. There is a double variety.

RANUNCULUS

Cultivation. Crock 5 in. pots well and add a handful of sedge peat and then a handful of silver sand. Plant 'bulbs' with claws pointing downwards in top layer of sand. Put the pots into a frame, but do not cover with sand or peat. Bring into the greenhouse in mid-January at a temperature of 50°F and stand on the staging. Plant in autumn for spring flowering.

SCILLA (Siberian Squill)

Cultivation. Treat as chionodoxa or muscari.

Species and varieties

S. bifolia – tropical sky-blue
S. bifolia rosea – pale pink
S. siberica Spring Beauty – china-blue

TRITONIA Beautiful, brightly coloured S. African plants, very popular with the ardent flower arranger.

Cultivation. Treat as chionodoxa (p. 150).

Varieties. A good one to grow is the rich deep orange Orange Delight.

TULIP The earliest tulips being of the Duc van Thol varieties are naturally very dwarf, only 3 or 4 in. high. Tulips have the usual typical bulb structure with swollen basal leaves surrounded by brown scale leaves. A great variety of colours may be obtained, from white to very deep reds, with bright yellows, oranges and even darker hues also available.

Cultivation. For the earliest flowers it is necessary to pot the bulbs in August, only the top-sized ones being used. The potting and covering with soil or sand outside are as for narcissus (p. 155), except that for the earlier varieties the sand or peat depth need only be 4 in.

The bulbs should be left alone in the covered-in beds to encourage root action till the end of November. Then lift the first batches and place under the greenhouse staging. It is now necessary to force them into flowering by darkening the area under the staging and maintaining a temperature of about 60°F. After five days the temperature should be raised to 65–70°F. The bulbs should be kept in the dark at this heat until the flower is well developed. They must then be moved into the light and put on the staging and the temperature reduced to 60°F for the foliage to green up and the full flower colour to develop. For later batches, the bulbs need not be forced in the dark but merely moved into the houses from outside and grown on at a steady temperature.

Varieties

SINGLE EARLY
Bellona – yellow; delicious scent
Brilliant Star – large scarlet flower, not too tall
Flamingo – soft cherry-rose with soft pink inside
Mon Tresor – short and sturdy, deep golden-yellow
DOUBLE EARLY
Golden Ducet – a fully double yellow, lasts a long time
Electra – deep crimson-lake with pale margins
Murillo – pale rose petals with white base
Snow Queen – a well-shaped pure white
Scarlet Cardinal – a well-formed glowing scarlet on strong stems
TRIUMPH TULIPS
Dutch Princess – deep glowing orange, petals edged with gold
Edith Eddy – carmine-rose edged with white
Golden Show – shining buttercup, broad petals
Crater – a crimson with dusky bloom
Princess Beatrix – a dusky orange-red touched with gold
DARWIN TULIPS
Clara Butt – salmon-pink
Flying Dutchman – cherry scarlet, very impressive
Golden Age – buttercup yellow, golden orange at edges

Normal Culture

AGAPANTHUS (African Lily) The leaves are long, narrow and evergreen. The tubular flowers are blue or white, and borne in umbels of 30 or more.

Cultivation. This is in flower in May and June, and should be potted in the Levington or Alexpeat compost No. 3. Bulbs are large, and need potting firmly and so that they are three-quarters covered. Grow in large pots or even tubs. Feed with Marinure once a week in the summer and water freely in the summer, but keep almost dry in the winter. Re-pot in March once every seven years; grow in a temperature of 50°F from March to September, and 35°F from September to March.

Species

A. africanus – deep violet-blue flowers on 20 in. stems
A. orientalis albus – white flowers

AMARYLLIS or HIPPEASTRUM This makes a nice specimen pot plant for the late spring and summer. The leaves are green and lanceolate, arising from a large onion-like bulb. The flowers, which grow upon long fleshy stems, are crimson, white and mottled. It has often been classed as a 'near evergreen'.

Cultivation. The hippeastrum is propagated by means of seeds or by offsets.

Seed. When propagating by seed, it must be realised that the plants will not be fit for flowering for three years. It is best to sow the seed as soon as it is ripe, usually about August, in standard seed trays containing the usual seed compost. During this time the greenhouse temperature should be maintained at 65–70°F.

When the plants have reached the third leaf stage, they should be pricked out into further boxes, 6 × 6 in., in a compost consisting of 1 part loam, 2 parts sedge peat and 1 part sand. A temperature of 55–60°F should be maintained and the young plants frequently syringed. As soon as the plants have developed a bulb about the size of a cherry they should be potted on into 3 in. pots using Levington or John Innes potting compost No. 1.

When the plants are potted, only one-third of the bulb should be buried below the compost, which should be made fairly firm. The pots should be set out upon the staging, allowing sufficient room for the leaves not to touch during growth.

Offsets. Offsets may be removed from the stock plants during spring, and potted into small pots as described above for seedlings.

As soon as the plants show – by a gradual slowing down of growth – that they require potting on, they should be moved into 8 in. pots using Alexpeat or Levington compost No. 3.

The plants are grown on steadily for two years, and during the autumn of the second year they should be fed well once a fortnight with any good liquid compound fertiliser. After this feeding-up process the hippeastrums may be partially dried off.

Watering. Only a very little water is necessary during winter: just enough to keep the plants alive.

1. *Ficus elastica* 'Tricolor', p. 191
2. *Freesia × kewensis* Mixed Hybrids, p. 15
3. *Gerbera jamesonii* (Farnell's Strain), p. 129
4. *Hedera helix lutzii*, p. 174
5. *Hippeastrum equestre*, p. 160
6. *Ixora*, p. 110

Temperature. A temperature of 45–50°F should be maintained. Large hippeastrum bulbs may be urged into growth at any time between February and September by plunging the pots into moist peat at a temperature of about 60°F. The plants should be watered regularly and fed every week with a good liquid compound fertiliser, and under this treatment they should flower in five weeks. It may be necessary to shade the house in really hot weather to prevent the sun bleaching the flowers.

Ripening off. After flowering, the plants should be placed in a cool greenhouse, and fed so as to 'fatten up' the bulbs. As soon as the leaves begin to turn yellow, less water should be given. Any bulbous offsets may be potted up singly during spring.

Varieties. There are no varieties as such; but it is possible to get exhibition hybrids in various colours, e.g. red, crimson, orange, salmon-pink, pure white and striped. These bulbs are on the dear side, and may cost £1 or more each. *A. belladona purpurea* is much cheaper.

CLIVIA (Kaffir Lily) Evergreen leaves and tubular flowers in shades of scarlet, orange and yellow.

Cultivation. The plants are in bloom from January to July, and are best grown in Levington compost or Alexpeat compost No. 3. It isn't necessary to re-pot each year; I usually delay it for six or seven years and then do the work in February because I find the plants do best when pot-bound.

The temperature from March to September should be 60°F, but from September to March it need only be 45°F. Propagation is by division of roots in February. The bulbs should be planted in large pots with their tips peeping out of the compost. They should be watered well in spring and summer, but only sparingly in the autumn and winter.

Species

- *C. gardenii* – an orange yellow
- *C. kervensis* – canary yellow
- *C. miniata* – scarlet and yellow, has a number of good varieties

CRINUM (Cape Lily) Lily-like flowers borne in pairs on stems 2–3 ft. high, in bloom from April to October.

Cultivation. Grow the bulbs in the Levington or Alexpeat compost No. 2; re-pot every three to four years in March.

The temperature from March to September should be 55°F, and from September to March 45°F. It is possible to sow seeds in the spring at a temperature of 65°F, or to propagate by removing offsets from the sides of bulbs in March. Water well in spring and summer, but sparingly in autumn and winter. Lay the pots on their sides from November to January, and then bring them into a sunny part of the greenhouse in February.

Species

C. macowanii – white and purple
C. moorei – white and red

CYCLAMEN (Sowbread) The leaves are often marbled and the flowers may be white, pink, crimson, salmon and scarlet; sometimes the petals are fringed. They can be in bloom from October to March.

Cultivation. The corms can be potted up in July and August each year using the Levington or Alexpeat compost No. 2, or they can be left for two years in the same compost (the corm should always be partly above the surface of the compost). The plants should be grown in the greenhouse at a temperature of 50–55°F.

The corms can be raised from seeds sown at any time from August to November or even from January to March, and the little seedlings that arise are potted on into 3 in. pots. The plants need watering moderately until growth begins, and then regularly until flowering ceases. They should be kept on the dry side from May to July, and it is usual to lie the pots on their side during the resting time.

Species. C. latifolium, and its many hybrids.

EUCHARIS (Amazon Lily) The leaves are evergreen and the flowers large, white and sweet scented. In bloom from December to March.

Cultivation. The potting compost should be either the Levington or the Alexpeat compost No. 2, and it is only necessary to re-pot every three or four years in June or July.

The temperature of the greenhouse should be 70°F from March to September, 55°F from October to December and 65°F from January to March. Offsets may be removed from the big bulbs in June or July and potted up into 3 in. pots, or in February or March seeds may be sown ½ in. deep in Levington compost at a temperature of 85°F.

Give plenty of water in the spring and summer, but water only moderately for the rest of the year. Syringe the plants over each day on hot summer days, and after the second year top-dress the pots each March with ½ in. more compost.

Species. E. grandiflora – the stems are 1–2 ft. tall.

HAEMANTHUS (Red Cape Tulip, Blood Lily)

Cultivation. In bloom in spring and summer, and does best in the Levington or Alexpeat compost No. 2. Re-potting should be done every four years, and the bulbs should be buried to half their depth. The temperature should be 50°F in September to March, and 60°F from March to September.

It is possible to propagate by offsets taken off the sides of the bulbs in September or October in the case of the spring-flowering species, and in March and April in the case of the summer-flowering species. Give very little water until the bulbs start to grow, and then water fairly freely. It is a good idea to apply Marinure once a week when flowering. When the flowers fade, water sparingly, and later stop it altogether to allow the bulbs to dry off.

Species

H. albiflos – white, flowers in September
H. katharinae – red, flowers in April
H. puniceus – orangey red, flowers in August

LACHENALIA (Cape Cowslip)

Cultivation. Pot up in August in Alexpeat compost No. 3, placing bulbs only ½ in. deep in a 6 in. plastic pot. Water well and put in a frame. Leave till mid-November, and then put the pots on shelving near the light in the greenhouse at a temperature of 50°F. When growth starts, watering must be moderate, but when the plants are doing well it can be increased. When flower-spikes appear apply Marinure once a week, and when the flowers are over stop feeding and reduce water until hardly any is given. Then ripen the bulbs by placing pots outside in a sunny spot on a gravel path or a concrete yard. The plants must rest from June to the end of September, after which they may be started off again as before in November.

Species

L. tricolor – the red and yellow Leopard Lily
L. orchioides – yellow tinged red
L. pendula – a pendulous variety
Nurserymen often offer named hybrids as well

LEUCOCORYNE Another member of the lily family, a 'cousin' of the amaryllis.

Cultivation. Plant in September in 6 in. plastic pots filled with Alexpeat compost No. 1. Put five bulbs per pot, burying them 3 in. deep. Plunge the potted bulbs in sedge peat in a frame and water moderately. When growth starts bring them into the greenhouse and put on the staging at a temperature of 40°F, increasing to 50°F in early March. Should be in flower during March.

Species

L. ixioides odorata – 'Glory of the Sun'

NERINE A beautiful S. African member of the amaryllis family with clusters of iridescent white, pink or red flowers.

Cultivation. Pot up in August or September, three bulbs to a 6 in. plastic or clay pot, and plant to half their depth. After soaking in water put pot in a cold frame in the sun. Bring into the greenhouse in October and put on the staging at a temperature of 50°F. Water moderately from September to May, when flowers should show (the flowers always appear before the leaves). After flowering, reduce water until compost is quite dry, and keep it so from June to September. When growth is seen again in September, soak the pots well and start into growth. Can be grown in the same soil and pots for four years.

Species

N. sarniensis, The Guernsey lily – large umbels of flowers, various colours
N. bowdenii – pink
Nurserymen offer other species and varieties

VALLOTA (Scarborough Lily) Does not lose its leaves in winter like the amaryllis. Strap-like foliage and numerous scarlet flowers borne on 18 in. stems, blooming in late summer and autumn.

Cultivation. Pot up bulbs in autumn in Alexpeat compost. Water liberally in summer, but sparingly in winter. Does not need re-potting for 3 or 4 years.

Species

V. speciosa (syn. *V. purpurea*) – easy to grow if you don't fuss it

CHAPTER 16

Shrubs, Climbers and Twiners

Shrubs

THERE are a number of quite ordinary garden shrubs which can be made to flower early by being grown in the greenhouse, adding beauty to the glasshouse and the home. The plants are not regarded as permanent residents under glass; they come in for a certain period and are then 'forced', speeding growth and flowering. Don't just dig up any old shrub from the garden – it will probably be too large anyway and may not be suitable. Go to a reliable nurseryman (I am always willing to suggest names) and buy small compact specimens which have been grown primarily for this somewhat specialised forcing. By the correct pruning, by keeping the shrubs growing in small pots, by any necessary training and by the use of the correct compost, the nurseryman can supply you with a plant at a reasonable price which can be forced the first year, then rested out-of-doors, forced the third year, and then rested again, and so on.

It is usual to buy small potted shrubs in November or early December, though the order can of course be placed several months beforehand, thereby ensuring getting the best specimens. The shrubs should arrive potted really hard in a well crocked 6 in. pot. Straight away plunge the pot and contents in a bucket of water for an hour so that it is thoroughly soaked, and then put the pot in deep sedge peat or sand out-of-doors in a sheltered, semi-shady spot on a path or concrete yard.

To get the best results the shrubs must have fully ripened wood. This means that it should grow in the open in its pot during the summer and stay in the open during the cold, wet, fog and snow, until it is brought into the greenhouse to flower.

Forcing. The word 'forcing' gives the wrong impression to the beginner, who may feel he must put the potted shrub into an almost red-hot greenhouse full of steamy atmosphere. This is of course an exaggeration – and it is far from the case. This method of bringing into growth *must* be gradual, and ideally entails bringing the shrub into the greenhouse at a temperature of 40°F, and 10 days later increasing to 50°F, and maybe 14 days after that increasing to 70°F. This is impossible to do in a mixed house, and all the gardener can do is to keep the temperature of the greenhouse as low as possible in an attempt to suit all the plants growing at the time. And if you bring the pots into the greenhouse about the middle of December there is every chance that the specimens will be in bloom by late February, depending on the species and variety, of course. Anyway, it is quite a good plan to have several pots so that one can be brought in during December, another during January, a third during February, and so on. Thus you maintain a continuous show of flowering shrubs over quite a long period.

Watering and syringing. In addition to any heat that may be given in order to bring the shrubs into flower it is necessary to see that the pots are kept watered. The roots must never be allowed to become dry or there will be bud-dropping and no flowers. So water every day or every other day as necessary – more water is needed when the foliage is well out than when it is starting to grow.

In addition, syringing helps tremendously. If possible use water of the same temperature as that of the house. I stand a bucket or two in the house for this purpose so that the water is always warmish. Syringe at about 10 a.m. and again at 4 p.m.

Which Shrubs to Grow

The easiest of all the shrubs to grow in this way are probably forsythias, deutzias, lilacs, spiraeas, flowering currants, and a prunus such as *Prunus triloba multiplex*, which must be cut back hard after flowering. There are, however, a fairly large number of others that may be grown in this way, and a list of them appears below with certain special instructions as to their management. Remember that in all cases the idea is to grow smallish specimens as naturally as possible out-of-doors, and only when the right time comes bring them into the warmth to flower.

ACACIA (Mimosa) A beautiful Australian shrub and tree with yellow flowers, in bloom from January to March.

Cultivation. 2 in. cuttings of semi-ripe wood may be taken in summer, and these should be struck in the greenhouse propagator in a potting compost consisting of 4 parts soil, 1 part sedge peat and ½ part silver sand. Bring into the greenhouse in October two years later.

Species

- *A. armata*, the Kangaroo Thorn, fluffy flowers, grows easily, flowers in January
- *A. drummondii* – yellow flowers, March
- *A. longifolia*, Sydney Golden Wattle – yellow flowers, March

ACER (Maple) Grown principally for its glorious red and gold foliage.

Cultivation. Pot in any good garden soil, and grow in the greenhouse at a temperature of 60–65°F. Be sure not to spray the plants over when the leaves start developing, or the globules of moisture may cause scorching of the leaves on hot sunny days. Plant out in the garden after their use in the greenhouse ends.

Varieties of the Japanese maple *A. palmatum*:

A. p. dissectum atropurpureum – crimson leaves
A. p. ornatum – bronzy-red leaves
A. p. septemlobum elegans – purple leaves
A. p. dissectum aureum – golden leaves

ASTILBE (Spiraea) Really a herbaceous perennial. White, pink and red flowers in February.

Cultivation. Pot in any garden soil, making sure the roots are firmly packed. Bring into the greenhouse in January and grow at 60°F. Grows very quickly and soon produces flowers.

Species

A. japonica – white
A. arendsii – white, pink or red
A. davidii – rose pink

AZALEA Azaleas belong to the genus *Rhododendron*, blooming from early December to late April. Easy to force into flower.

Cultivation. Propagate in summer by cuttings struck in Alex sedge peat or Levington compost, and put in greenhouse propagator. Nurserymen often graft named kinds on common azalea stocks. Needs syringing over every day, and regular watering. Grow out of doors in partial shade from early June to late September. Remove any seed pods that form. Prune each year by shortening straggly growth.

Temperature. 50°F for *R. mollis*, 60–65°F for others.

Species

R. indica, the Indian Azalea – evergreen
R. mollis – deciduous. Flowers in March before new leaves appear
R. japonica – bring into greenhouse in December

CEANOTHUS (Californian Lilac) Sun lovers, blooming in April. They must never be forced to flower early, so do not bring into the greenhouse until February – when left late they flower better and the colour is deeper and brighter.

Cultivation. Buy sturdy specimens in 6 in. pots in November and put them up to their rims in peat in a frame until February. They hate being potted on, so keep in the same pot for years. Propagate by cuttings 3 in. long in peat and sand in September.

Species

- *C. dentatus* – blue evergreen
- *C. azureus* – deep blue, panicles often 6 in. long
- *C. parvifolius* – semi-evergreen, pale to deep blue flowers

DEUTZIA (Japanese Snowflower) In bloom in March and April. Often planted out as a border shrub after one year's forcing.

Cultivation. Propagate by 3 in. long shoots in June in any compost and put in propagating frame. Pot up in November. Keep out of doors from July to November. Water sparingly from October to late February and freely from April to late September. Prune early June after flowering, cutting back the shoots that have blossomed. The temperature should be 60–65°F.

Species and varieties

- *D. gracilis* – beautiful white flowers
- *D. g. aurea* – with yellow leaves
- *D. g. wilsonii* – white; really a hybrid

FORSYTHIA (Golden Bells) Magnificent golden flowers in February, before the leaves.

Cultivation. Very easy to force. Keep in the greenhouse from December to April, and it will respond to a temperature of only 50°F. Water sparingly until it comes into flower, and then freely. Prune back hard after flowering and keep the bushes compact. Plunge pot outside in sheltered spot from the end of May until December.

Species

- *F. intermedia spectabilis* – numerous rich coloured golden flowers
- *F. densiflora* – very fine golden yellow flowers
- *F. viridissima* – a lovely Chinese variety

HAMAMELIS (Wych-Hazel) A deciduous shrub flowering in January.

Cultivation. Buy compact little bushes in November, preferably in 6 in. pots. Plunge in sedge peat in a frame, water carefully, bring into the greenhouse in mid-December and put on staging. Propagate by layering in November into any potting compost. The greenhouse temperature should be 60°F.

Species

H. mollis – fragrant yellow flowers, reddish-bronze foliage
H. japonica zuccariniana – lemon-yellow flowers

KERRIA (Jews Mallow) In bloom in March, April and even early May.

Cultivation. Pot up in any compost in October, or buy ready potted up in November, and bring into the greenhouse at 55°F in January. Water moderately all the time as this plant particularly hates soddenness. Propagate by taking cuttings 3 in. long of young shoots and put into a sandy peat compost in the propagating frame.

After flowering and pruning back, place the shrub in a sunny spot and feed with a fish manure – about 1 oz. per pot – and water in. Should be ready to use again next winter.

Species

K. japonica, sometimes called *Corchorus japonicus* – yellow
K. j. pleniflora – a lovely double yellow

LABURNUM (Golden Chain) Blooms at the end of March.

Cultivation. Dwarf shrubs should be bought in November. Bring into the greenhouse in early February, water regularly and syringe over every day. Propagate by layering in October or November – any compost is suitable. Temperature should be 55°F. Can be budded onto seedling stocks.

Species and varieties

L. anagyroides – yellow flowers
L. a. aureum – golden flowers and leaves
L. watereri – long semi-weeping racemes

PRUNUS (Flowering Peach) One of the prettiest of the forced shrubs and fairly easy to grow. Flowers in February and March.

Cultivation. Bring into the greenhouse in late November or early December and provide a temperature of 60°F for early flowering, or 40°F for the March flowering. Any compost is suitable.

Species and varieties

P. persica – pale pink
P. p. alba plena – lovely double white
P. p. sanguinia plena – double carmine flowers
Nurserymen may also offer additional named varieties

RIBES (Flowering Currant) Flowering in early March.

Cultivation. Bring into the greenhouse in December, and provide a temperature of 50–55°F. Syringe over regularly in February. Propagate by 6 in. long cuttings in sandy soil in October out of doors.

Species and varieties

R. sanguineum – rose-coloured
R. splendens – deep red
R. album – white
R. Pulborough Scarlet – deep red flowers

SYRINGA (Lilac) Flowering February to March.

Cultivation. Bring into the greenhouse in November at a temperature of 60°F. Syringe daily but water moderately. Increase the temperature to 65°F when the buds burst, reducing to 55°F when they open. After flowering, prune back shoots to within 2 in. of their bases. Keep in the greenhouse until about 20th May, then plunge pots in deep sedge peat out of doors. They must rest for two years before being forced again. They should be potted in a good loam.

Species and varieties. The fragrant common lilac is *S. vulgaris*, of which there are a number of named varieties.

Mrs Edward Harding – a semi-double, scented, red
Sensation – purple-red edged with white
Edward J. Gardner – semi-double light pink
Katherine Havemayer – deep purple-lavender, gorgeous scent

VIBURNUM (Snowball Tree, Guelder Rose) Flowers in February.

Cultivation. Propagate by 3 in. long cuttings of half-ripened shoots in July, using any loamy soil. The greenhouse temperature should be 60°F. The plants can be forced for two years running, and can remain in the same 6 in. pots for this period.

Species and varieties

V. carlesii – white, delicious scent
V. carlesii aurora – red buds and pale pink flowers, delicious scent
V. c. × *burkwoodii* – evergreen, fragrant white flowers 4 in. across
V. c. × *burkwoodii*, Park Farm Hybrid – similar to above, but with somewhat larger flower-heads
V. c. × *judii* – pink-tinted scented flowers

Climbers and Twiners

A large number of climbers and twiners can be grown in the greenhouse and it is possible with these attractive plants to clothe the back wall of a lean-to house, train them along the eaves of the normal span-roofed house or, in the bigger house, to grow them up the purlin posts. Climbers may have to be helped by tying them up from time to time, but the twiners will twist round any support, and manage to scramble up on their own.

BOUGAINVILLEA Flowering all summer. The 'flowers' are in fact coloured bracts surrounding insignificant little yellow flowers.

Cultivation. Pot in Levington or Alexpeat compost No. 3, but it really does best when growing in the greenhouse soil. Rest during winter, keeping compost or soil almost dry, and start watering in early March. Take 3 in. long cuttings of young shoots in March or April and strike in the propagator at 70°F. Prune hard each winter cutting back all laterals to within 1 in. of their bases. Grow in a sunny spot at 45–50°F, and feed regularly with liquid manure in summer.

Species and varieties. B. spectabilis is the most common species, growing to about 15 ft. It has several named varieties:

Lady Wilson – rosy-cerise
Mrs Butt – bright rose
Russell's Orange – orange bracts

CALATHEA A foliage plant grown for the beauty of its leaves.

Cultivation. Pot in Levington or Alexpeat compost No. 2, and grow in the shadier part of the greenhouse at a summer temperature of 70°F and a winter one of 65°F. Re-pot fairly firmly in March. Water well from May to the end of September and moderately from then on. Propagate by division of the roots in March.

Species

- *C. backemiana* – tuberous roots and longish silvery-grey leaves with emerald-green blotches
- *C. lindeniana* – oval leaves 6 in. long and 3 in. wide, dark green with emerald-green area, underside maroon
- *C. louisae* – leaves 7 in. long, 3 in. wide; medium green with variegated areas, underside greenish-purple
- *C. mackoyana* – one of the most beautiful foliage plants: leaves green and silvery-green, underside greenish-purple
- *C. zebrina* – emerald-green leaves with darker green stripes and dark purple and greenish-purple undersides

CESTRUM (Bastard Jasmine) A climber flowering in June and July. Some cestrum species have been known as habrothamnums, and may be found as such in a few catalogues.

Cultivation. Pot in Levington or Alexpeat compost No. 3 or plant in the greenhouse soil. Temperature September to March should be 40–45°F and from March to September 45–60°F. Re-potting may be necessary and should be done in March. Prune back each February to keep in good shape. Propagate by taking 3 in. long cuttings of side shoots in July or August, and strike in a sandy soil in a propagating frame at a temperature of 65–70°F.

Species

- *C. aurantiacum* – orange-yellow
- *C. newellii* – bright crimson
- *C. fasciculatum* – purple-red
- *C. psittacinum* – bright orange

CISSUS (Kangaroo Vine) An evergreen growing to 8 ft. and said to be the most beautiful foliage plant which can be grown in the greenhouse. This is a true climber and need not be tied.

Cultivation. Pot in Levington or Alexpeat compost No. 3 or plant in the greenhouse soil. From March to October keep the temperature at 45°F, and from October to March at 40°F. Likes plenty of light and lots of water in summer. Give liquid manure every two weeks in summer. Propagate by 2 in. long cuttings in sandy compost in the propagator at 80°F.

Species

- *C. antarctica* – grows freely; oak-like leaves
- *C. capensis*, Cape of Good Hope Vine – hardy and vigorous, with leaves

rather like those of the grape vine. Young leaves pink and hairy, older ones with autumn tints before falling. Train up a trellis

C. discolor – very beautiful. Leaves a vivid metallic green marbled with white and purple, shaded with crimson and peach; underside a deep crimson. Regular watering imperative and syringing over in the growing season. Hates a draught

CLERODENDRON (Glory Tree) Flowering in summer, this is an easy plant to grow.

Cultivation. Propagate by 3 in. long cuttings in January or February, struck in a sandy compost in the propagator at 70°F. Pot in Levington or Alexpeat compost No. 3, or plant in the greenhouse border at the warmest end. The October to February temperature should be 55°F, and from February to October 65–70°F. Must be partially rested in winter. They require a moist atmosphere and should be given little water in winter, but once growth has begun in February, give increasingly more.

Species

C. thomsonae – crimson and white flowers. There is a variegated form, and a variety *balfori* with larger flowers

C. splendens – crimson flowers in June and July

C. fragrans – scented rose and white flowers, downy shoots, deciduous

C. pleniflorum – white suffused with pink

COBEA SCANDENS (Cups and Saucers) Flowering in summer, this is really a greenhouse climbing perennial, but more usually grown as an annual.

Cultivation. Sow seeds in March ¼ in. deep in 6 in. pots of Levington or Alexpeat compost No. 3 at 60°F, or in the greenhouse border. Likes a cool airy spot and a temperature of about 50°F.

Species. C. scandens – bell-shaped mauve flowers. A very swift climber.

FICUS (Climbing Fig) An evergreen with ivy-like growth, producing aerial roots which support the dainty, climbing branches.

Cultivation. Pot in Levington or Alexpeat compost No. 3. Do not overwater in winter, as the leaves will turn yellow, but do not ever allow the soil to dry out. Give Marinure regularly in summer and provide a temperature of 50–60°F.

Species and varieties

F. pumila – rather like ivy, with small heart-shaped leaves. Will not stand direct sunlight. One or two variegated varieties available

HEDERA (Ivy) A comparatively easy evergreen to grow.

Cultivation. Pot in Levington or Alexpeat compost No. 2, in a 3 in. pot, and when the roots completely fill this, pot on to prevent it becoming pot-bound. Propagate by cuttings taken in May and strike in the propagating frame – they take a long time to root.

Species and varieties

H. helix Pittsborough – dark green leaves with lighter veins, 1¼ in. long and 1 in. wide
H. h. Chicago – large, brighter green leaves
H. h. Chicago variegated – a beautiful type
H. h. minima – the smallest, thinnest-leaved ivy
H. h. Maple Queen – very tiny leaves, vigorous grower
H. h. Golden Jubilee – small golden leaves, slow grower. Almost the only variety which prefers sun to shade
H. h. cristata – leaf edges fringed like parsley
H. h. Lutzii variegated – leaves flecked with cream and white spots
H. h. Adam – a similar but improved type
H. h. Eva – a further sport with smaller leaves
H. h. Glacier variegated – grey green leaves, margins variegated
H. canariensis can stay four years in a 3 in. pot. Hates over-watering

HOYA (Wax Flower)
H. bella and *H. carnosa* flower in summer; *H. australis* in October.

Cultivation. Pot in Levington or Alexpeat compost No. 3. *H. bella* should be kept at 50°F, the other two being quite happy at 40°F. Shoots may be layered in sandy peat in summer or propagated in March or April by cuttings of the previous year's growth, struck in sandy compost in the propagating frame at 75°F.

Species and varieties

H. australis – scented waxy flowers, dark green leaves
H. bella – small white waxy flowers, which are highly scented. Should be grown along the roof ridge so that the blooms, which hang down, can be seen.
H. carnosa variegata – fleshy oblong leaves and flesh-pink waxy flowers.

Plant may send up leafless shoots which look bare for a while, but always eventually produce leaves

H. c. variegata aurea – golden yellow leaves with dark green margins

IPOMOEA (Morning Glory) An easy-to-grow, sun-loving plant with beautiful large flowers all summer.

Cultivation. Propagate in March by seeds sown ⅛ in. deep in Levington compost at a temperature of 60°F. Although it can be started in a 3 in. pot it must be moved to a 6 in. or 8 in. pot (Levington or Alexpeat compost No. 3) for later growth. Treat as an annual, discarding the plants at the end of the season. Water well – often twice a day – and syringe once a day in summer. Provide a temperature of 50°F.

Species

I. cardinalis – cardinal red
I. hybrida Darling – wine red with a white throat
I. h. Pearly Gates – pure white
I. rubro-caerulea praecox – beautiful sky blue, numerous large flowers. Outstanding

JASMINUM (Jasmine) Vigorous, easy-to-grow plant, with scented flowers from February to May.

Cultivation. Pot in Levington or Alexpeat compost No. 3 in a 6 in. pot and allow the stem to twine naturally. Pinch out the growing points once a month from April to October to prevent untidiness and elongation. When flowering is over, cut back the plant by half. Grow in a nice sunny spot at 60°F. Propagate in April and May by cuttings of firm shoots in a propagating frame at 65°F.

Species

J. polyanthum – very fragrant flowers, white inside and rosy outside
J. primulinum – bright yellow flowers in March and April

LAPAGERIA (Chilean Bellflower) Bell-shaped waxy flowers in summer.

Cultivation. Pot in Levington or Alexpeat compost No. 3, and re-pot once into 8 in. pots in February or March. Train the shoots up the wall and along the rafters of the greenhouse if desired. Prefers a shady spot. Temperature should be 40°F from October to March and 55°F from March to October; ventilate freely all summer. Requires watering thoroughly from May to September and daily syringing from March until the flowers appear. Propagate by seed sown on sandy peat at 70°F in March, or layering in sandy peat in autumn.

Species

L. rosea – rosy crimson flowers, faintly spotted
L. r. albiflora – white flowers
L. ilsemanni – larger, more brightly coloured flowers

OPLISMENUS (Basket Grass) The hanging stems make this a suitable plant for the front of the greenhouse staging.

Cultivation. Pot in Levington or Alexpeat compost No. 1. Re-potting is seldom necessary. Water freely in summer, moderately in autumn and winter, and give Marinure once a fortnight in summer. Temperature: 50–55°. It propagates easily by forming roots at the joints.

Species

O. hirtellus variegatus – trailing stems and pretty lance-shaped leaves with ivory and green stripes. Hardy

PASSIFLORA (Passion Flower) A very striking and easy plant to grow, flowering in summer.

Cultivation. Pot in Levington or Alexpeat compost No. 3 or plant in the greenhouse border, where it probably does best. Re-pot into large pots in February or March. Train shoots up wires or posts. Water plentifully and syringe over daily from April to September, water moderately at other times. Feed with Marinure once a week when in flower. Prune in February, shortening strong shoots by one-third and weak ones hard back. Root pruning increases the quality and quantity of flowers.

Species and varieties

P. antioqulensis – rich rose-red flowers
P. caerulea – vigorous, white and blue, slightly scented flowers
P. c. Constance Elliott – white
P. racemosa caerulea – rosy crimson
P. suberosa – greenish-yellow flowers
P. watsoniana – purple stems, white and violet flowers

ORCHIDS. *Above, Cymbidium* 'Rio Rita' x 'Radiant'. *Below, Odontoglossum* 'Pengelly' x 'Dolcoath'.

Above, Primula obconica 'Sutton's Giant Blue'. *Below, Nemesia strumosa* 'Orange King'.

PHILODENDRON

Cultivation. Pot in Levington or Alexpeat compost No. 2, and re-pot in February or March. Water all year round and syringe every day. Keep all plants out of draughts and the temperature at 70°F from April to the end of September, and 65°F from October to March. Propagation is by 1 in. long cuttings taken at any time of the year and struck in sandy peat in the propagating frame at a temperature of 75°F.

Species

P. tuscla – large arrow-shaped shiny leaves
P. imbe – vigorous climber, bright green arrow-head leaves. Likes warm conditions
P. manei – train upright. Heart-shaped leaves with a wrinkled surface, dark green mottled with silver
P. micans – slender climber or trailer, purplish leaf
P. pertusum – smallish leaves. Has quite a climbing habit
P. soderii – strong climber with large leaves. Likes warmth
P. hastatum and *P. burgundy* are also worth trying

PITTOSPORUM (Parchment Bark) A New Zealand plant which will grow out of doors in a very sheltered spot in the south-west.

Cultivation. Pot in Levington or Alexpeat compost No. 3, and re-pot in March or April. The greenhouse temperature should be 40° in winter and 60°F in summer. Water moderately in winter and spring, freely in summer, syringing twice a week in summer. Propagate by 2 in. long cuttings of firm shoots, struck in the propagating frame at 55°F. Prune straggling shoots in September.

Species

P. eugenioides – very attractive and prolific foliage

RHOICISSUS (Climbing Pomegranate) A fairly good evergreen climber, originally from Natal, South Africa.

Cultivation. This plant is somewhat more tender than Cissus (p. 172), and the general cultivation conditions should be the same.

Species and varieties

R. rhomboidea – shiny dark green toothed leaves, requiring sponging from time to time. Young growth brownish
R. rhomboidea Jubilee – larger leaves, often $4\frac{1}{2} \times 2\frac{3}{4}$ in.

ROSA (Climbing Rose) Blooming from March to May.

Cultivation. Plant in the greenhouse border in October/November or in 18 in. × 2 ft. tubs of Levington compost. Give bone meal once a year, at 4 oz. per square yard. Water freely from March to end of October, syringe over daily in spring and summer, and keep fairly dry in winter. Feed with Marinure from April to August. Paint the branches with a tar oil wash such as Mortegg in November, using a 5% solution. Prune each winter, cutting out some old wood and tying in the new. Be sure to open the ventilators well in autumn to let the wood ripen properly.

Varieties. Two good old-fashioned ones are:

Marechal Niel – golden yellow flowers
Niphetos – white

SAXIFRAGA (Tumbling Saxifrage) A very easy plant to grow, with marbled or variegated leaves.

Cultivation. Pot in Levington or Alexpeat compost No. 1 – re-potting is seldom necessary. Water moderately in summer, a little in winter, and provide a temperature of 45–50°F. Propagate by runners.

Species

S. stolonifera (syn. *S. sarmentosa*) – produces runners like strawberry plants, and is easy to propagate. Has hairy marbled leaves and likes a light but sun-less spot
S. tricolor – a more difficult, variegated species

SCINDAPSUS (Giant Leaf) An evergreen producing larger and larger leaves as it grows older.

Cultivation. Start off in a 6 in. pot, preferably in Levington compost No. 2, and do not re-pot. Never over-water as the plants rot at the leaf joints if too wet. Syringe over in summer and provide a temperature of 60–65°F. Do not grow it in direct sunlight. Propagate by short cuttings.

Species and varieties

S. aureus – oval dark green leaves flecked with yellow. Prefers to grow up bark or moss, or even a live branch
S. a. Golden Queen – yellow leaves
S. a. Marble Queen – white-netted leaves
S. pictus argyraeus – heart-shaped leaves 2 in. long and olive-green dotted with silver. Keep moist during growing season. More delicate than *aureus*

SOLANUM (Climbing Potato, Potato Vine) Flowering in August.

Cultivation. Plant in March in any well-drained soil or in Alexpeat compost in tubs. The October to March temperature should be 55–60°F, from April to September 67–70°F. Water freely in summer and moderately in winter. Propagate in late August or early September by cuttings of young shoots struck in a sandy peat compost in the propagating frame with bottom heat.

Species

- *S. wendlandii* – lilac-blue flowers 2½ in. across
- *S. jasminoides* – bluish-white flowers ¾ in. across. Will grow in a temperature of 40°F
- *S. j. grandiflorum* – a more robust variety
- *S. j. variegatum* – blotched creamy-white leaves
- *S. pensila* – bright violet-blue, star-shaped flowers, followed by pale violet fruits rather like cherries

STEPHANOTIS (Madagascar Jasmine) Flowering in spring and summer, with long-lasting, fragrant, white tubular flowers and fleshy, firm and shiny leaves.

Cultivation. Pot in Levington or Alexpeat compost No. 2 and re-pot into 6 in. pots in February or March. Train the shoots up wires or to the rafters, but shade from the sun. Water well from April to the end of September, but moderately for the rest of the year. Syringe daily from April to the end of September, but not when the plant is actually in bloom. Temperature should be 65–70°F. Propagate in spring by cuttings of the previous year's wood, struck in the propagating frame at a temperature of 70°F with bottom heat. Watch out for mealy bug and scale, and spray with Malathion.

Species

- *S. floribunda* – a twiner with pure white scented blooms in large bunches. At its best in May
- *S. thouarsii* – a more slender plant, with white flowers in groups of three

STREPTOSOLEN A climbing hairy evergreen shrub, native of Columbia, flowering in June.

Cultivation. Pot in Levington or Alexpeat compost No. 3 or in the greenhouse border, where it does well. Re-pot if necessary in March or April. The March to October temperature should be 60°F, and 50°F for the rest of the year. Water well in summer, also feeding with weak liquid manure once a fortnight, and moderately in winter. Immediately after flowering, prune all the shoots moderately. Take 2 in. or 3 in. long cuttings in spring or early summer and strike in the propagator at a temperature of 60°F.

Species. S. jamesonii – bunches of orange bell-shaped flowers.

THUNBERGIA Flowers in summer. Some plants bear leathery fruits with sword-shaped beaks.

Cultivation. Pot in Levington or Alexpeat compost No. 3 or plant in the greenhouse border. Re-pot into 5 in. pots in May or June. Propagate by seed in March or April sown $\frac{1}{16}$ in. deep in Levington compost at 75°F, or in March to May by 3 in. cuttings of young shoots, struck in peat and sand in the greenhouse propagator at 75°F. Water well in summer and syringe over with considerable force to prevent thrip damage. After a few years, prune back bare and unsightly shoots.

Species

- *T. alata* – dark purple and cream flowers. Up to 6 ft. tall
- *T. alata* hybrids – coral, salmon, buff and orange colours, with a black zone around the throat
- *T. angulata* – pale blue flowers, yellow throats
- *T. coccinea* – red flowers in pendant racemes
- *T. fragrans* – white scented flowers

TRACHELOSPERMUM (Chinese Jasmine) A somewhat tender climber but easy to grow and train, flowering in July and August, and worth growing for its beautiful scent.

Cultivation. Pot in Levington or Alexpeat compost No. 3, or plant in the greenhouse border. Re-pot into 8 in. pots or tubs. Water moderately in summer, but less in winter, and feed weekly with Liquinure in June and July. Temperature should be 55–60°F. Propagate in July or early August by 3 in. long ripened shoots, struck in the propagator with bottom heat.

Species and varieties

- *T. asiaticum* – evergreen, with clusters of yellowish-white scented flowers
- *T. jasminoides* – evergreen, very fragrant white flowers about 1 in. across
- *T. j. variegatum* – similar to above, but with silvery variegated leaves

CHAPTER 17

Foliage Plants

THERE is a tremendous vogue – almost a craze – for house plants these days, and it is claimed (and there is little doubt that it is true) that there are probably three or four times the number of house plants in Great Britain as actual people in the country. The hobby probably started in Scandinavia and spread, as so many things do, to the United States, reaching the United Kingdom some 15 years ago.

The man with the greenhouse has, therefore, a tremendous opportunity for raising and growing his own foliage plants and ferns. He can not only grow these plants the more perfectly in the greenhouse, but may take them into his home for a period of time if he wishes, and then bring them back into the greenhouse for propagation, resuscitation, or even just a rest.

Foliage plants prefer to grow in the shadier parts of the greenhouse, and, where amateurs are able to devote a whole house to such plants, it is better to provide shading from about the middle of May until the middle of October in the south and south-west, and from early in June until the end of September in the north. A simple way of providing such shade is to stretch a piece of hessian of the right length and width right the way across the sides of the greenhouse and to fix it at either end with temporary nails or large round-headed carpet-tacks.

Generally speaking, most plants need watering regularly in the summer but only occasionally in the winter. The larger-leaved plants transpire (lose moisture) more and so need a greater amount of water than those plants with very tiny leaves.

A lot of foolishness has been talked about the high temperatures most pot plants need because they are said to come from the tropics. Actually, of course, many of the plants come from the mountainous regions where night temperatures may fall to 45°F, and where there is a high rainfall and therefore a moist atmosphere. For this reason, generally speaking, a greenhouse for foliage plants should be syringed all over from time to time so as to provide this moist atmosphere. Any form of heating tends to dry the air in a greenhouse, but the syringing over of the main path and the gravel or Lytag on which the pot plants are growing will help prevent this happening.

Foliage plants grow well in plastic pots because these do not lose moisture through their sides, as do clay pots. The smaller plants can be accommodated in 3 in. pots, most of the larger ones in 6 in. pots, and very large specimens in 8 in. pots.

Temperatures. There are delicate foliage plants that must have a temperature above 50°F and there are tender plants where the temperature must not fall below 60°F, but the great bulk of foliage plants will grow well at a temperature of 45°F in the winter, and won't grumble even if the temperature falls to 40°F.

The famous firm of Thomas Rochford, which sells thousands of foliage plants, always labels its specimens according to how difficult they are to grow. A blue label is for the intermediate type of plant, a pink label for the easy-to-grow plants, and a yellow label for the more difficult ones – a considerable help when buying a previously untried plant.

Watering. The look of a plant often enables a gardener to decide how much water is needed. A plant like a cactus, with thick fleshy succulent leaves, is able to put up with prolonged dry periods whereas plants with soft thin leaves will suffer immediately if water is withheld. At periods when a plant is not growing, hardly any water is needed; in fact if you saturate the compost in the pot at that time you merely encourage bacteria and fungus diseases, and root rot may occur. This is one of the reasons why watering must be kept down to a minimum in the wintertime.

In the south it is usually necessary to start watering regularly and seriously about the beginning of May, but in the north, where the winters are longer, spring watering may be postponed until the second or third week of May. After the winter period of minimum watering it is quite a good idea in the spring to stand the pot plant in a bucket of water to give it a thorough soaking. After this it can be stood on the staging of the greenhouse, and when it starts to grow water it again.

Feeding. Never attempt to feed a plant until it has started to grow. It is the roots of the plants that are to take in the feeds you give, and unless these are vigorous and happy, they cannot do so. Wait, therefore, until it is certain that the roots are growing again and then feed with a liquid manure, which only has to be dissolved in water before application. If the gardener wishes to inspect the roots to make certain that they are growing, all he has to do is invert the pot with his fingers spread over the top of the compost and with the other hand to tap the edge of the pot on the edge of the staging and the whole of the soil ball should come out undisturbed. It is easy to see if there are new roots by looking for the white tips.

Feeding should only be done in the summer, and should always cease by the end of September. In fact it is a good plan to stop feeding in the autumn when there is no sign of any new leaves developing.

Potting on. Though this has been mentioned in a general way on p. 67, it is as well to explain that for foliage plants the work is best done sometime during the spring and summer, i.e. while the plants are growing well. Such plants have a good root action which will quickly penetrate the new compost around the ball of soil in the bigger pot. This new compost must always be firmed well with a potting stick so that it is firmer than the compost in the ball. When new clay pots are used at potting on time, these have to be soaked in water for two hours or so before being used.

When the Alexpeat compost is used this need not be firmed more than with the thumbs all around the ball, and at the end of potting on the surface should be level. It helps if the pot is tapped on the bench from time to time during the re-potting process. After potting on give the plants a good watering.

Pinching out. In order to prevent plants from growing too tall for the small greenhouse, with a pair of scissors or a sharp knife, or even with the thumb and forefinger, cut or pinch out the growing points of the plants. In many cases this will cause the specimen to grow bushier. The side-shoots which develop may be pinched back in their turn and interesting-looking plants produced by this method.

Plants such as the ivies can often be cut back harder than this, as also can a plant like *Philodendron scandens.* This pinching back is usually done in late April because by then the roots are active again.

Of the very large number of foliage plants which can be grown in a greenhouse, a careful selection has been made of plants that are both easy to grow and particularly popular. The list is by no means complete, and those who are anxious to grow specimens not mentioned in this chapter should consult the Royal Horticultural Society's Dictionary of Gardening in its five big volumes.

ACALYPHA (Copper Leaf)

Cultivation. Propagate in March by 2 in. cuttings struck in sand and peat in the propagator at 80°F. Pot in a compost of equal parts sedge peat, loam and silver sand, re-potting in March. Grow on the greenhouse staging at a temperature of 70–75°F from March to September, and 60°F for the rest of the year. Water fairly well in summer, little in winter. Thin out in February.

Species

A. godseffiana – 2 ft. tall, green leaves with yellow margins
A. wilkesiana – coppery-green leaves mottled with red

ACORUS (Sweet Rush) Seldom grows higher than 12 in.

Cultivation. Propagate by division in April. As this is a marshy plant the pots must never be allowed to dry out completely, and it is hardly possible to over-water. Pot in any heavy loamy soil, re-potting in March. Temperature: 40–45°F.

Species

A. gramineus variegatus – thin variegated green and cream leaves

ALOCASIA Likes heat, and is not really suitable for mixed house

Cultivation. Propagate in spring by division of the rhizomes. Pot in Levington compost mixed with some chopped sphagnum moss, and re-pot in March. Grow at 65°F from September to March occasionally watering, and 75°F March to September watering well.

Species

A. sanderiana – glossy green leaves with white margins
A. indica metallica – olive green leaves with a metallic sheen

ARAUCARIA (Norfolk Island Pine) An evergreen plant popular with the Victorians.

Cultivation. Propagate by seed sown in the greenhouse at 65°F or by 3 in. cuttings of soft wood in September. Pot in Levington or Alexpeat compost No. 1 and grow in a sunny spot if possible, at 55–60°F, although it will also grow in the shade. Feed once a month in summer. Re-pot only occasionally, as this plant grows better when not disturbed.

Species

A. excelsa, the true Norfolk Island Pine
A. glauca – slightly more silvery foliage than *excelsa*

ASPARAGUS (Asparagus Fern) Very decorative feathery leaves.

Cultivation. Propagate by division of roots in March, or by seed at 65°F, also in March. Pot in Levington or Alexpeat compost No. 2, re-potting in March. They like partial shade (and may be grown under the greenhouse staging if desired), plenty of air and a temperature of 60°F. Really well-established plants may be fed weekly, phosphates being very important for healthy growth. Syringe over daily in summer.

Species and varieties

A. plumosus – will climb if allowed
A. p. nanus – dwarf
A. sprengeri – thin sharp leaves, often used in hanging baskets

ASPIDISTRA (Parlour Palm) An old-fashioned plant now coming back into popularity.

Cultivation. Pot in any compost, re-potting rarely, in March. Keep out of strong sunlight, and water freely in summer, moderately in winter. Temperature: 45°F.

Species and varieties

A. elatior (= *A. lurida*) – dark green shiny leaves
A. e. variegata – leaves with alternate stripes of green and white

BEGONIA Very useful pot plants, with variegated and coloured leaves.

Cultivation. Propagate by leaf cuttings in spring or summer, by stem cuttings in spring, or by seed sown in February at 70°F. Pot in Levington or Alexpeat compost No. 3, re-potting in March. Grow at 60°F and feed every few days during the growing season. Water well in summer, moderately in winter. Maintain a moist moving atmosphere and keep plants away from gas fires, which they hate.

Species and varieties

B. masoniana – the Iron Crop begonia
B. rex hybrids – green leaves marked with pink, grey and red
B. ricinifolia – bronze-green leaves. Grow in shade
Many good varieties such as Silver Queen, Isolde, Heligoland and Our Queen.

CAREX (Blue Grass)

Cultivation. Propagate by division of roots in March. Pot in Levington or Alexpeat compost No. 1, and re-pot in March every three or four years. Grow at 50°F, in some shade, and water freely in summer, moderately in winter.

Species

- *C. morrowii variegata* – stiff grass-like leaves 4 in. long with a cream stripe along the edges. (Is often sold as *C. japonica.*)

CHLOROPHYTUM One of the oldest house plants. Useful in positions where the small plantlets which develop after the flowers can hang decoratively down.

Cultivation. Propagate by division in spring or from the little plantlets produced by the plant. Pot in Levington or Alexpeat compost No. 2, re-pot in spring. Grow at 60°F and feed with Marinure once a week in summer. Water freely in summer and sparingly in winter.

Species

- *C. comosum variegatum*, Spider Plant – long narrow green- and ivory-striped leaves
- *C. orchidiastrum* – dark bronzy-green leaves with copper-coloured stems. Usually propagated by seed

CITRUS (Orange) Although it does bear small white scented flowers, this is included in this section as it is always grown for its leaves and fruit.

Cultivation. Propagate by 3 in. cuttings in July. Pot in Levington or Alexpeat compost No. 2, and re-pot in April. Water freely in spring and summer, syringing over daily in summer and moderately in winter. Temperature: 55°F. Prune from time to time to promote bushy growth. Some have been lucky enough ro raise good plants from seed.

Species

- *C. sinensis* – a slow grower, with glossy green leaves and small fruits like miniature oranges
- *C. mitis* – the beautiful Calamondia orange

CODIAEUM (Croton, South Sea Laurel) Many different types are available, with green leaves marked with red, mauve and yellow.

Cultivation. Propagate at any time, dipping the cuttings in charcoal to stop 'bleeding'. Pot in as small pots as possible, in Levington or Alexpeat compost No. 2, seldom repotting. To get the best colours in the leaves, grow in full sun, in moist conditions, at 60°F, and take great care to avoid sudden changes of temperature, as this causes the leaves to fall.

If the plant loses its leaves in winter, cut it down to within 6 in. of the top of the pot.

Species and varieties

C. variegatum pictum – grass-like green leaves spotted with orange
C. Van Ostensee – very attractive and worth trying
Newer types include *C. v.* Philip Geduldig; *C. v.* Hollufiana; *C. v.* Mrs Iceton; *C. v.* Julietta

CORDYLINE

Cultivation. Propagate at any time by offsets inserted in 2 in. pots of sandy soil, or in March by seed sown in the propagator with bottom heat. The potting compost may be Levington or Alexpeat No. 2, but re-pot very seldom, as cordyline does best in small pots. Never grow in direct sunlight, and maintain a temperature of 55°F from March to September, and 45°F for the rest of the year. Water well in summer, moderately in winter.

Species and varieties

C. terminalis – long leaves curiously striped with red and green
Juno – the leaves may be cerise-pink, flecked, or purplish-brown marked with red
Firebrand – bright cerise leaves with dark purple and cerise margins
Margaret Story – mid-green leaves with pink edges
Rededge – a very striking type

CYANOTIS (Cushion Plant)

Cultivation. Propagate at any time by 2 in. cuttings struck in the propagating frame. Pot in Levington or Alexpeat compost No. 2, re-potting seldom. Grow at 64°F, in the sun in summer and in a little shade in winter. Water freely in summer, moderately in winter.

Species

C. moluccensis – somewhat like tradescantia, leaves slightly hairy, and green with a central purple fleck

CYPERUS (Galingale)

Cultivation. Propagate by division of roots in March or by seeds sown at 55°F in February. Pot in Levington or Alexpeat compost No. 3, and seldom re-pot. Temperature: 60°F March to September, 45°F September to March. Feed once a week in summer when growing well, and add water to a dish in which the pot stands, so moisture is drawn up from the bottom of the pot.

Species

- *C. diffusus* – grass-like leaves, easy to grow
- *C. alternifolius*, Umbrella Plant – numerous long narrow leaves at the top of a tallish stem

DICHORISANDRA (Seersucker Plant)

Cultivation. Propagate by division in March. Pot in Levington or Alexpeat compost No. 2, re-potting in March. Temperature: 75°F from March to October, 60°F for the rest of the year. Water freely from March to the end of September, moderately from October to February.

Species and varieties

- *D. reginae* – a beautiful, slow-growing plant with 5 in. long leaves which are blackish-purple and silver on top, dark purple beneath
- *D. mosaica* – dark green oval leaves covered with thin white lines; purplish beneath. It produces bright blue flowers, which I cut off
- *D. m. undata* – popular wavy-leaved and compact plant, with silvery- and blackish-green leaves, purple beneath

DIEFFENBACHIA (Dumb Cane) An extremely poisonous evergreen with large, usually variegated leaves.

Cultivation. Propagate by 3 in. stem section cuttings at 80°F, first removing the growing tip of the stem. Pot in Levington or Alexpeat compost No. 2 and re-pot in March. Water freely in summer, moderately in winter, syringing daily in summer. Temperature: 60°F (never allow it to fall below 50°F). Grow in partial shade.

Species

- *D. imperialis* – large leaves, mid-green with cream markings, 12 in. long and about 4 in. across
- *D. oerstedii* – dark green leaves with a central ivory stripe, usually 9 in. long and 4 in. across
- *D. picta memoria* – silvery-green leaves with irregular green margins
- *D. exotica* Perfection – green spotted oval leaves on a cream background
- *D. amoena* – large green oval leaves, creamy white along the veins
- *D. rhoensi* – oval leaved, creamy white, slow grower

DIZYGOTHECA An elegant, very slow-growing member of the Aralia family.

Cultivation. Propagate by cuttings of side shoots in March. Pot in Levington or Alexpeat compost No. 2, making sure there is very good drainage. Re-pot in February or March, but never over-pot. Grow in sunlight, at 70°F from March to September, 60°F September to March. Water freely from April to the end of September, moderately in winter.

Species

D. elegantissima – thin spidery leaves, often divided into ten sections. Young foliage is brownish-red, turning nearly black later. Subject to severe attack by red spider unless the atmosphere is kept moist

DRACAENA (Dragon Plant) A very popular group of plants for indoor decoration.

Cultivation. Propagate in March by 1 in. cuttings of side shoots, partially buried horizontally in a sandy peat compost in the propagating frame. Alternatively, propagate at any time by offsets. Pot in Levington or Alexpeat compost No. 3, re-potting in February or March in Alexpeat compost No. 3. Grow at 50–55°F, water freely in summer, syringing daily, and moderately in winter. Feed once a week in summer when growing well. If the plants get too big, cut them down quite hard. The cut-off part can be potted in a 3 in. pot – it will usually grow.

Species

D. deremensis – palm-like, with long leaves striped with silver and dark green
D. fragrans – long green and gold leaves, up to 18 in. and 4 in. across
D. f. lindinii – leaves with central gold band and wide gold edges
D. godseffiana, Florida Beauty – dark green oval leaves thickly spotted with cream
D. sanderi – thin leaves 1 in. across, with undulating edges of ivory, centres greyish-green
D. marginata concinna tricolor – a very beautiful tri-coloured type

ELETTARIA (Ginger Plant) Rather like a small bamboo in appearance, with leaves 6 in. long, 1½ in. across and scented with a cinnamon-like perfume.

Cultivation. Propagate by division of the rhizomes in March. Pot in Levington or Alexpeat compost No. 2 and re-pot in summer each year.

Grow at 60°F in partial shade, as strong light discolours the leaves. Apply Marinure once a week in summer. Water freely from March to September, moderately from September to December and leave nearly dry from December to March.

Species

E. cardamonum – downy scented green leaves

EUCALYPTUS (Australian Gum Tree)

Cultivation. Propagate in April or August by seeds grown at 65°F. Pot in Levington or Alexpeat compost No. 3, re-potting frequently, in spring. Grow in a sunny position at 60°F, water freely in summer and slightly in winter. Can be planted outside after a year or two in the greenhouse.

Species

E. citriodroa – citrus-scented gum
E. globulus – blue gum with slightly scented leaves

× **FATSHEDERA** The '×' prefix indicates that this is a cross between two genera, i.e. *Fatsia* and *Hedera*.

Cultivation. Propagate by cuttings at a leaf joint, struck in the propagating frame at 75°F. Pot in Levington or Alexpeat compost No. 2, seldom re-potting. Grow in a cool spot at 50°F, water well from April to August and moderately for the rest of the year.

Species and varieties

× *Fatshedera lizei* – evergreen semi-climber with 5-lobed dark green leaves
× *F. l. variegata* – wide cream margins to the young leaves, becoming narrower with age

FATSIA (Fig Leaf Palm) Sometimes known as *Aralia sieboldii*, this is an easy and very attractive plant to grow.

Cultivation. Propagate by seed at 65°F. Pot in Levington or Alexpeat compost No. 2 and seldom re-pot; if this should be necessary, February or March is the best month. Grow at 50°F, in a fairly shady spot, water well in summer and seldom in autumn and winter.

Species

F. japonica – dark green leaves with 7–9 lobes. Will grow well in a 6 in. pot. There is also a variegated kind

FERNS This heading covers an enormous number of genera and species, too large to describe here. Personally, I do not find them as exciting as other foliage plants.

Cultivation. Propagate by division of the plants in March. (It is possible, though difficult, to propagate by sowing spores at 65°F.) Pot in Levington or Alexpeat compost No. 2 and re-pot in spring. If re-potting back into the same pot, trim back the roots first. Grow at 50–55°F and keep the atmosphere moist in summer. Water well in summer, moderately in winter, using rain water if the tap water is alkaline. Syringe over during hot sunny weather. Feed with Marinure once a month in summer when the plants are growing well. Never over-pot.

Species

Adiantum cuneatum, Maidenhair Fern – does not like sun; do not syringe over
Asplenium bulbiferum, Mother Spleenwort – pretty dark green fronds
Dryopteris linnaeana, Oak Fern – very decorative fronds
Osmunda cinnamonia, Cinnamon Fern
Phyllitis scolopendrium, Hart's Tongue Fern
Polystichum acrostichoides, Christmas Fern
Pteris cretica – dark green fronds on slender stems

FICUS

Cultivation. It is advisable to leave the propagating of this plant to the nurseryman – buy it ready-potted. As it much prefers to grow in as small a pot as possible, e.g. a 5 ft. plant does best in a 6 in. pot, re-pot only if necessary, in early May, using Levington or Alexpeat compost No. 3. Grow at 55–60°F and keep on the dry side in winter, starting to water carefully again in spring – over-watering causes the leaves to turn yellow. The leaves of ficus species need cleaning once a fortnight.

Species

F. elastica, India-rubber Plant – evergreen, leathery green leaves up to 12 in. long, 7 in. wide
F. e. decora – the variety most often grown. Larger leaves, red sheath over the growing point
F. e. doescheri – variegated leaves edged with yellow and with lighter green markings. Delicate

F. benjamina – very different from the other species; narrow pale green leaves growing in a weeping habit. Likes more water than the other kinds

F. indica – 'leggy', with 4½ in. long, shining leaves

F. lyrata, Banjo Fig – thick leathery leaves 12 in. long, 9 in. across

F. pumila – trailing or climbing, heart-shaped leaves about 1 in. long. Must be well shaded

Other species that are well worth growing include *F. robusta*; *F. benghalensis*; *F. radicans variegata*; *F. Pandui formis* and *F. diversifolia*

FITTONIA A Peruvian plant named after two botanists, Elizabeth and Sarah Fitton.

Cultivation. Propagate in March by cuttings of firm shoots struck at 75°F – they take easily as a rule. Pot in Levington or Alexpeat compost No. 2 and re-pot every four years in March. Grow at 60°F, always in the shade. Water freely in summer, syringing every three days, moderately in winter.

Species

F. argyroneura – oval or nearly circular leaves, 4 in. long, with silvery veins. A trailer

F. verschaffeltii – more elongated leaves, with a network of carmine-coloured veins

GREVILLEA (Silk Bark Oak) Popular for its silvery fern-like leaves.

Cultivation. Propagate by seed sown ¼ in. deep in No-Soil compost at 65°F, or by 3 in. cuttings of stem with a small piece of old wood attached, struck in the propagating frame in April. Prefers to grow in the sun, but will tolerate partial shade. Temperature: 60°F March to October, 50°F for the rest of the year. Water freely in summer but keep rather dry in the winter. Do not prune unless absolutely necessary.

Species

G. robusta – very easy to grow, producing a lovely plant within six months of sowing seed. Very prone to attack by red spider

HELXINE A small close-growing Corsican plant with tender green very shiny leaves.

Cultivation. Propagate by division in April. Grow at 50°F, never in bright sun – it will grow under the greenhouse staging or hang over

Above, a display of small cacti. *Below, Lithops,* the strange 'pebble plant'.

CLIMBERS. *Above left*, the white-flowered form of *Lapageria rosea; Right*, Morning Glory *(Ipomoea rubro-caerulea)*. *Below*, Bougainvillea 'Golden Glow'.

the edges of pots. Try always to water from underneath, i.e. by standing the pot in a dish of water, and never let the pots dry out. Pot in Levington or Alexpeat compost No. 2. Re-potting is seldom necessary.

Species

H. solierolii – tiny green leaves, moss-like habit

LIGULARIA (Leopard Plant) Large, usually toothed or lobed leaves; the flowers should be removed as soon as they appear.

Cultivation. Propagate by division of the plants and pot in Levington or Alexpeat compost No. 3. Re-pot only when necessary, into 6 in. pots. Grow at 55–60°F, preferably in the shade. Water freely in summer, less in winter, but never allow the pots to dry out.

Species

L. tussilaginea – 18 in. tall
L. aureo maculata – golden leaves
L. argentea – silvery leaves

MARANTA (Arrowroot Plant)

Cultivation. Propagate by division in March. Pot in Levington or Alexpeat compost No. 2, re-potting every year in March. Prefers to grow in the shade, at 60°F from March to October, 55°F October to March. Keep almost dry in winter, give a little water in autumn and plenty in spring and summer, syringing every day in summer to keep the atmosphere moist. Try to rest the plants during the winter months. The best results will be had if the plants are left in small pots.

Species and varieties

M. leuconeura – light green leaves with pale mid-ribs
M. l. kerchoveana – low-growing habit, with oval leaves 5 in. long and 3½ in. wide. Young leaves have red blotches on the veins, later turning maroon
M. l. massangeana – Leaves 4½ in. by 2½ in., white veins with maroon areas between, later becoming dark green. Underside of the leaves a rosy-purple colour
M. l. erythrophylla – known as the Red Herringbone Maranta

MONSTERA (Shingle Plant) Large erect plants with aerial roots and much dissected large leaves.

Cultivation. Propagate any time by 2 in. cuttings struck in sandy peat at 70°F. Pot in Levington or Alexpeat compost No. 3, re-potting in March. Likes to grow against a damp wall. Temperature: 70°F in summer, 60°F in winter. Water freely from April to mid-October syringing twice a day, moderately in winter syringing once a day.

Species

M. deliciosa – too large for the small greenhouse, with leaves up to 4 ft. long

M. pertusum (syn. *M. adamsonii*) – leaves perforated by slits and holes, 12 in. long and more-or-less heart-shaped

PANDANUS (Screw Pine) As the plants age, the trunks become twisted, giving rise to the common name. The word pine derives not from the tree, but from the pineapple, whose leaves are similar.

Cultivation. Propagate from suckers in March. Pot in Levington or Alexpeat compost No. 1, re-potting in February or March. Grow at 55°F, in the sun, keep the plants in small pots. Water freely in summer syringing regularly to maintain a moist atmosphere, but keep on the dry side in winter. Never feed.

Species

P. candelabrum variegatum – leaves 3–4 in. long, 2–3 in. across, bright green with longitudinal white bands. Only suitable for the large greenhouse

P. sanderi – variegated leaves, slightly spiney and banded with yellow

P. veitchii – dark green leaves bordered with silver

PELLIONIA A member of the nettle family, and distantly related to the hop.

Cultivation. As for pilea, p. 195.

Species

P. daveauana – brownish bronze-green leaves 2 in. long. Keep well pruned as may otherwise become bare in the centre. Like the Artillery plant this can emit clouds of pollen

P. pulchra – rounded silvery-green leaves 1 in. long, with dark green veins. Never goes bare in the centre, like *daveauana*, but is rather tender. Keep on the dry side in winter

PEPEROMIA (Pepper Elder) Although included here as primarily a foliage plant, this does have a rather interesting flower, looking somewhat like a mouse's tail.

Cultivation. Propagate by cuttings or a single joint with a leaf attached in March, in the propagating frame at 70°F with bottom heat. Pot in Levington or Alexpeat compost No. 2, seldom re-potting. Grow at 45–50°F in a shady part of the house, always keeping out of draughts. Water well in summer, slightly in winter.

Species

- P. *peperata* – a mass of small heart-shaped leaves which are corrugated, purplish and dark green
- P. *hederaefolia* – heart-shaped wavy-surfaced leaves 2½ in. long and 2 in. across, pale grey with olive-green main veins
- P. *obtusifolia* – leaves 2 in. long and 2 in. wide, dark green with a purple edge. Do not over-water
- P. *argyreia* – leaves marked with alternate bands of silver and dark green, and dark red stalks. Hates draughts
- P. *magnoliaefolia variegata* – leaves 2 in. long, 1¾ in. across. When young they are grey-green in the centre with cream edges, which later become light green

PHILODENDRON Other easy to grow species in this genus are climbers, see p. 177.

Cultivation. Propagate any time by 1 in. cuttings struck in sand and peat in the propagating frame at 75°F. Pot in Levington or Alexpeat compost No. 2, re-potting in February or March. Temperature: 70°F April to the end of September, 65°F for the rest of the year. Water well all year round, and syringe every day.

Species

- P. *pertusum* – smaller leaves, plus a climbing habit
- P. *scandens* – though naturally a climber, the shoots can be pinched back to produce a bushy plant. Leaves heart-shaped, 4 in. long, 2½ in. across
- P. *elegans* – also naturally a climber. Leaves 15 in. long and 12 in. across
- P. *bipinnatifidum* – a palm-like plant with heart-shaped young leaves. When re-potting, take care not to break the fleshy roots
- P. *tuxla* – large arrow shaped shiny leaves

PILEA (Artillery Plant, Gunpowder Plant) When the pot is tapped on the bench the plant produces masses of pollen, hence the common name.

Cultivation. Propagation extremely easy by cuttings struck at 60–65°F in February or March. Pot in Levington or Alexpeat compost No. 2, re-potting into small pots in March. Grow at 60°F in partial shade. Water sparingly in autumn and winter, freely in summer. Requires regular pruning if it is not to become leggy and floppy.

Species

P. cadierei – oval leaves 3 in. long, 1½ in. across, dark green with silver areas between the veins. A quick grower
P. microphylla – tiny leaves arranged in a rosette, and therefore sometimes called Mossy Pilea
P. mollis – called for obvious reasons the Moon Valley Pilea

RHOEO Very similar to tradescantia; does well in hanging baskets.

Cultivation. Propagate by offshoots from the bottom of the stem, used as cuttings. Pot in Levington or Alexpeat compost No. 2, re-potting in February or March. Prefers to grow in the shade, at 40°F from October to March, and 55°F April to October. Do not allow the compost to become too wet in winter as this causes the plants to rot.

Species and varieties

R. discolor – long thin fleshy leaves, somewhat like an aloe, dark green above and rosy-purple beneath
R. d. vittatum – vivid green stripes on the upper surface of the leaves

SANSEVERIA (Bow-String Hemp) Popular house plants with erect leathery leaves; tolerant of drought.

Cultivation. Propagate by division of plants in March or April. Pot in Levington or Alexpeat compost No. 1, re-potting in March. Grow at 55°F, on the sunny side of the house. Water little in summer and only once a month in winter. Feed with Marinure once a week in summer. Hates excessive watering, particularly in winter.

Species

S. hahnii – rather like an aloe, with mottled dark and grey-green leaves
S. cylindrica – cylindrical pointed leaves in two shades of green
S. trifasciato laurentii – known as the Mother-in-Law's Tongue

SELAGINELLA Dainty moss-like plants.

Cultivation. Propagate in summer by cuttings struck at 70°F – mist propagation is very effective. Pot in Levington or Alexpeat compost No.

1. *Narcissus bulbicodium* 'Hamilton'
2. *Jacobinia carnea*, p. 110
3. *Primula malacoides* mixed, p. 115
4. *Pilea cadierei*, p. 195
5. *Ruellia macrantha*, p. 117
6. *Plumbago indica* (*P. rosea*)

1, re-potting in February or March, keeping the plants in small pots. Water freely and syringe over regularly in summer. Temperature: 50–55°F. Throw away the plants at the end of three years and start again.

Species

- *S. emmelinia* – branching and erect, 1 ft. tall
- *S. lepidophylla*, Resurrection Plant – so called because it curls up when dry and opens out again when watered
- *S. martensii* – 9–12 in. tall, evergreen and moss-like. There is also a variegated kind
- *S. kraussinia* – a creeper often used in hanging baskets

SETCREASEA (Purple Heart) A very fascinating plant to grow.

Cultivation. As tradescantia, see below.

Species

- *S. striata* – leaves 2 in. long and 1 in. wide, dark green striped with ivory and purplish underneath
- *S. purpurea* – erect oblong leaves and stems of tyrian purple. Cut back regularly as a very quick grower. Produces magenta flowers which I always cut off as they look horrible against the beautiful leaves and stems

SYNGONIUM (Goosefoot Plant) Similar in habit to the philodendrons. Its common name derives from the leaves, which are shaped rather like the webbed feet of geese.

Cultivation. Propagate by cuttings struck in the propagating frame at 70°F with bottom heat. Pot in Levington or Alexpeat compost, No. 2, seldom re-potting. Grow at 55–60°F, preferably in partial shade. Feed once a week with Marinure in summer, water sparingly in winter and early spring. Looks well if trained up moss-covered canes.

Species

- *S. podophyllum albolineatum* – green arrow-shaped leaves with white lines, the central one of which may be over 7 in. long and 2 in. across. The two side lobes are like arrow-heads
- *S. vellozianum* – a climber with dark green shiny leaves shaped like a three-headed spear

TRADESCANTIA (Spiderwort, Wandering Jew) A very popular plant, often confused with zebrina.

Cultivation. Propagate by cuttings in spring or summer. Pot in Levington or Alexpeat compost No. 2, re-potting every year. Water well in summer, sparingly in winter. Syringe once a week in summer. Temperature: 50–55°F.

Species

T. blossfeldiana – dark green leaves with purple undersides, and purple hair-covered stems
T. fluminensis – a very striking, firm growing type – a favourite of the Author's
T. fluviatilis variegata – green and yellow leaves with a pink tinge if the plant is grown in a sunny airy position and kept on the dry side. Often reverts to the plain green-leaved type. There is a golden variety whose leaves are larger

ZEBRINA (Zebra Plant) Easy to grow and very beautiful.

Cultivation. As for tradescantia, see p. 197.

Species and varieties

Z. pendula – leaves 2 in. long, 1½ in. wide and dark green with a purple stripe. Underside is green and rosy-purple. To get the best colours, keep on the dry side and in the semi-shade
Z. p. quadriacolor – leaves striped with white, rosy-purple, silvery-green and dark purple
Z. purpusii – the upperside of the leaf is dull purple, the underside vivid purple. Extremely easy to grow. Never over-water as this causes the purple colour to disappear. (Sometimes wrongly sold in shops as *Tradescantia purpurea*)

CHAPTER 18

Cacti and other Succulents

IT may be as well to start by explaining the difference between cacti and succulents, because although all cacti are succulents, not all succulents are cacti.

A succulent has either an extremely fleshy stem or very fleshy leaves, and sometimes both. The fleshy foliage is able to hold moisture during long hot sunny periods and thus enables the plant to grow in open dry situations. Very few succulents are hardy, and therefore they have to be given protection during winter. They are thus very suitable for the greenhouse subjects. Normal succulents include desert plants and those which grow in rocky conditions, e.g. the sedums, the agaves, the aloes and the mesembryanthemums.

Cacti on the whole have no leaves (with the exception perhaps of the *Pereskia*), so there is a minimum of transpiration. They usually have thorns or spines, and many of them produce flowers. Succulents, on the other hand, hardly ever bloom. The cactus belongs only to the family Cactaceae, but succulents are to be found in many differing plant families. There are, incidentally, 2,000 different species of cactus – divided into three main 'tribes', the Opuntieae, the Cereae, and the Pereskieae. (Though gardeners on the whole use the word 'succulent', this is not a term recognised by the expert botanist.)

The thorns on many types of cactus are there for three purposes – they encourage the collection of moisture, they give protection from feeding animals, and they deflect the sun's rays. In some cases the thorns are long and beautiful, and in other cases they are covered with what seem to be white hairs.

The colour of cacti blooms varies from violet to dark crimson, from red to pink and orange, and from white or yellow to purple. Most of the flowers have a beautiful metallic sheen to them. They are often produced at night-time and many last only a short time, some for only a few hours, others for two to five days. In the case of Stapelia, the odour is perfectly horrible, but in Echinopsis the perfume is delicious.

Flowers are usually followed by fruits, many of which are edible, like those of the Prickly Pear. It is claimed that the red fruits of one or two of the Mammillarias taste like strawberries.

Compost. The word compost is used here to denote a suitable mixture of soil, peat and sand in which the plants will grow satisfactorily. Descriptions of the various composts available will be found in Chapter 7.

The general aim should be to produce a compost which is open and porous. When a sticky clay soil has to be used it should be baked well first. It is inadvisable to use soil from the garden that has been well manured in the past, for this will be too full of organic matter. Really poor soil will do if mixed with an equal part of coarse silver sand, plus a teaspoonful of carbonate of lime for every 6 in. potful required.

Compost for cacti. Cacti like a very open compost. One which I use consists of:

1 part by bulk good heavy loam (sterilised)
1 part by bulk coarse silver sand (sterilised)
1 part by bulk finely broken or powdered burnt clay
$\frac{1}{16}$ granulated charcoal
$\frac{1}{32}$ ground clean eggshell

The Phyllocacti can do with a slightly richer compost, so add to the above one part of sedge peat.

Some keen cactus growers use a compost consisting of:

2 parts good loam
2 parts sharp silver sand
1 part small pieces of broken flower pots
$\frac{1}{2}$ part sedge peat

Add to each bushel of this mixture a 5 in. flowerpotful of bone-meal and a 5 in. flowerpotful of ground limestone.

Other experts make up a compost consisting of 1 part poor soil, 1 part medium grade sedge peat, and 1 part coarse river-washed sand. To this must be added (as mentioned before) 1 teaspoonful of carbonate of lime per 6 in. potful. This latter compost is particularly suitable for the epphitic types of cacti.

Some gardeners prefer to use powdered dried egg-shells instead of the lime and find this particularly good. Whichever compost is used, it must be passed through an $\frac{1}{8}$ in. sieve so as to get rid of all the dust. A cactus compost must never be too fine or else the seepage may block the drainage holes. The sand used in a cactus compost must always be very coarse.

John Innes potting compost No. 1 is sometimes used for cacti but the plant foods normally added for other plants must be reduced by a quarter. It helps also if 3 parts by bulk of coarse sand is used instead of the normal 2 parts. Cacti have also been grown quite successfully for two or three years in Levington compost. Incidentally this compost is particularly useful to those who live in flats, because it is light and ready to use, as is, of course, the Alexpeat compost.

Composts should be only just moist and never really wet. Never let anyone persuade you to use leaf-mould in a compost, because it may be full of diseases, weed seeds and insect pests.

Compost for succulents. A friend who specialises in succulents and doesn't grow the true cacti at all always uses a compost consisting of:

3 parts poor soil (sterilised)
2 parts medium-grade sedge peat
½ part brick dust
3 parts coarse-particled silver sand

Those who cannot get hold of brick dust may use fine, fully consumed coke ashes.

In all cases add a teaspoonful of dried crushed egg-shell per 6 in. potful and pass the whole through a ¼ in. sieve using, of course, the material collected in the sieve and not that which falls through.

Light. Cacti like to grow where there is plenty of sunshine, light and air. Give them also as much reflected light as possible by painting the walls and the staging with white gloss paint. Ideally the greenhouse should run north and south and be in a nice sunny place. It is always possible to grow ordinary pot plants on the staging and the cacti on the higher shelving near the glass. Thus the cacti get the maximum amount of sunshine and at the same time do not receive the drip from any other plants that are being watered.

Those who have a greenhouse in a town or city will know how quickly the glass gets fouled with greasy soot. It is necessary, therefore, to use a soft rag plus a good detergent in hot water, to wash the outside of the house twice a year, say in spring and autumn, and the inside of the house once a year, say in autumn. It is very important for successful cacti cultivation to allow the maximum amount of sunlight to get through the glass.

In cases where the gardener takes over a greenhouse which is not clean and has not been washed down for years, it may be necessary to remove the serious grime with fluoric acid, which may be obtained from the local chemist. It must be used extremely carefully because it burns into glass. It is put on with a brush and then must be hosed off with plain water one minute later, before the acid can etch the surface of the glass. While any washing down is being done, care must be taken to plug up the down spouts so as to prevent the detergent or acid from running into any tank that may be present to collect the rain water.

Watering. Fortunately cacti will put up with very bad treatment. Even when the pots are not watered for a long period, they will usually survive. It is only when the pots are over-watered and the soil is sodden that the roots rot and the plants die – there is seldom any need to water cacti in pots from the beginning of October until the beginning of March. When changing over from the no-water period to the regular watering in the summer, the gardener must start by giving very moderate quantities for the first four weeks, only after this may 'normal' watering be done. At the end of the watering season water must be gradually withheld over a period of six weeks.

It is a good thing to water early in the morning so that any excess moisture will have evaporated by night-time. In extremely hot summers watering may be done in the late evening after the sun has gone down. As far more harm occurs through over-watering than under-watering, the gardener must take care. The only exception is probably in the case of baby plants, which can be given a good soaking once a week. Large specimens may be watered once a day during the months of June, July and August.

Globe-shaped cacti need less water than the tall column types. Always give water if there is any sign of shrivelling, for this indicates that a cactus is living on its reserves. Some cactus growers prefer to dip rather than water, and they stand the pots up to their rims in a tank or sink for 5 minutes each day in the summer, and for only 2 minutes each day in the spring.

Heating and ventilation. In the Good Gardeners' Association Demonstration Greenhouses, the thermostat that controls the electric heating system is set at 48°F. It does not matter if in the summer the temperature rises to 60° or 70°F, when this rise is due to the sun and not to artificial heat. By using a thermostat the electricity is cut off the moment the sun heat rises above 48°F. It is of course possible to heat by means of warm water pipes or even fumeless paraffin stoves, and I have often saved my cactus plants from frost damage in an emergency by covering them over at night-time with several sheets of newspaper.

Cacti hate 'fug' so remember to open the ventilators of the greenhouse during the day when the weather is hot and dry, but close them when there is lots of rain about or any sign of fog. Automatic ventilators are ideal (p. 39). The purpose of ventilation is to give the plants fresh air and not to give them draughts. Even in the winter it may be advisable to open a ventilator a little so that a foetid atmosphere may get out and fresh air get in.

Special instructions for succulents. With succulents it is necessary to provide some shade during the hottest months of the year, either by means of automatic blinds, or by the use of a kind of green distemper known as Summer Cloud, which is sprayed over the outside of the greenhouse by means of a syringe.

Some succulent experts put a 6 in. layer of fully washed gravel or Lytag on the staging of the house and then sink the pots of succulents into this, so the containers are hidden. This plunging helps to prevent the pots drying out as well as protecting the roots.

Like cacti, succulents need plenty of light and air, and are watered in a similar manner. Though most plants want a little water in the winter, there are exceptions like the echeverias and aloes, which need watering once or twice every three weeks. The conophytums need no water from May until the end of July and the lithops from Christmas Day until the end of April. When buying plants it is a good thing to discuss watering with the nurseryman, it is then possible to put on the plant-labels somewhat cryptic figures such as 12–5, indicating to the gardener that no watering is to be done between December and May.

It will be seen that succulents need the same kind of treatment as cacti, and if any tying up is thought to be necessary, use a very thin stake dyed green by a Rentokil preservative, and then tied with green twist or twill.

Potting on. Transplanting or potting on is best done in March as a rule. If old pots are to be used again, they should be well scrubbed and, if they are made of clay, stood in cold water for two days afterwards. Where there has been any sign of disease put 1 teaspoon of permanganate of potash crystals in every 2 gallons of water used, and stir well.

Pot on into a pot slightly bigger than the previous one. Put on a good leather glove so that the plant may be held in one hand and tap the edge of the pot on the bench with the other. In this way the ball of roots and soil should come out of the pot cleanly. Carefully remove the broken crocks at the base of the ball of soil and with your fingers open out or free the basal roots a little. Now is the time to look carefully for the presence of pests and particularly the mealy bug, which looks like little spots of white cotton-wool substance on the roots. If this or any other insect is present, dip the ball of soil into diluted liquid nicotine for about 5 minutes. (To make this solution, dissolve a ¼ oz. of liquid nicotine, obtainable from the chemist, in a 2-gallon bucketful of warm water.) After the dipping the ball of soil must be dried out a little before potting on.

Now remove some of the old compost from among the massed roots, best done with a pair of tweezers. If any dead or matted roots are found, they may be cut out completely with a sharp razor-blade, and when roots are broken, these too should be cut back with a razor. The production of a new root system is encouraged if the ball is dusted after the root-pruning with a powdered hormone such as Seradix.

Put new crocks to a depth of 1 in. into the bottom of the larger pot, and see that the bottom crock is concave and rests over the hole. Over the crocks place 2 tablespoonfuls of medium-grade sedge peat, and over this place a ½ in. layer of the appropriate compost. Stand the ball of roots in the centre of the new pot, hold with one hand and scoop up the compost with the other hand so as to fill up the pot evenly and gradually all round the plant, turning the pot round as you do so. Tap the pot from time to time on the potting bench to help the compost to settle, but don't try to make it too firm. Leave about ¼ in. at the top of the pot to allow for watering later and unless you know your cacti very well indeed, label immediately.

Cacti

There is a wide choice of cacti that can be grown, and once bitten with the 'cactus bug', it is possible to collect some hundreds of species. It is impossible in a book like this to do more than suggest simple easy-to-grow species with which the beginner can make a start.

PERESKIEAE There are 19 species in this tribe, many of which are shrubby, leafy and vine-like. The leaves are large and fleshy.
Pereskia aculeata, the Lemon Vine – a rampant grower.
P. grandifolia – egg-shaped leaves about 6 in. long with almost black spines; one of the shrubby types.

OPUNTIEAE This group consists of eight genera and contains species which vary in size from quite small plants to quite large trees. The *Opuntia* is a genus with over 300 species and is commonly called Fig Cactus.
Opuntia cylindrica – has, as its name suggests, a cylindrical stem about 2½ in. in diameter, and bears rosy-red flowers 2½ in. wide.
O. ficus-indica, the true Fig Cactus from India – can have a stem 6 in. wide and covered with bluish bloom. There are usually no spines but the flowers are yellow and 4 in. wide.
O. polycantha – prostrate and spreading. The stems are 4 in. wide and usually wrinkled and warted. The flowers are yellow and 2½ in. across. There are a number of varieties, such as *rufispina*, with reddish spines, and *albispina* with white spines.
O. vulgaris is the Barberry Fig, which grows to a height of only 1 ft. The stems may be 1 in. wide and covered with greyish wool. The flowers are pale yellow and 2 in. across.

1. *Salpiglossis sinuata,* p. 90
2. *Sanseveria,* p. 196
3. *Schizanthus pinnatus* Mixed Giant, p. 90
4. *Sinningia speciosa,* p. 119
5. *Streptocarpus* × *hybridus* 'Constant Nymph', p. 121
6. *Thunbergia alata,* p. 92

CEREAE This is the largest tribe of cacti. Very few of its members have leaves and the stems vary considerably. The flowers are tube-like.
Aporocactus flagelliformis is the Rat Tail Cactus, which hangs over the edge of the pot with stems usually ¾ in. in diameter. The flowers are bright red and usually last through April and May.
Cephalocereus. This is a genus noted for plants with large columnar stems which may be branched. The plants often carry hair or wool, particularly at the tip, and the flowers are often hidden in this hair, appearing usually at night. The fruit is a small round berry.
C. senilis, the Old Man Cactus – so called because the plants are covered with soft baby white hairs each up to 5 in. long. The flowers are red outside, white within, and are often partially hidden at the top of the plant by the long hairs.
Cereus peruvianus, commonly known as the Hedge Cactus – tall and tree-like – the stems may become 8 in. thick and are bluish-green in colour, turning to dull green later. It has been known to grow to a height of 50 ft. The flowers open at night and are browny red outside and white inside.
Echinocactus. The plants in this genus have cylindrical stems often of considerable size. The ribs are straight and clearly marked. The flowers are usually yellow and borne on the tips of the plants, where they may be partially buried in the woolly hairs.
E. grusonii, the Gold Barrel Cactus – barrel-shaped and covered with

FIG. 24 Cacti: *left Mammillaria wildii; right Aporocactus flagelliformis*

yellowish wool. When really old it may be 4 ft. high. The flowers are brownish outside and yellowish inside.

Echinocereus. The stems of plants in this genus are short, globular, fleshy and soft, but often very prickly. The flowers are reddish-purple as a rule, large and showy and usually seen in the daytime.

E. delaitii – sometimes incorrectly called the Old Man Cactus because it is covered with long white hairs. The stems are yellowy green, and the plants usually grow to a height of 10 in. The flowers are pink and 3 in. long.

E. rigidissimus, usually called the Rainbow Cactus – red or brown, pink and white spines borne in alternating bands around the plant. Usually grows to a height of 12 in. and has purple flowers 3 in. long.

ECHINOPSIS The collective name for this particular genus is the sea urchin cactus because the plants are, on the whole, globular. It is an interesting genus because most of the plants bear sweet-smelling flowers.

E. leucanta – bears purplish-white flowers smelling like violets and grows to a height of 12 in. the stems being 5 in. across and grey-green in colour.

E. multiplex – a much branched type, only growing to a height of 6 in. Again the flowers are rose-coloured and very sweet-smelling.

EPIPHYLLUM This is known as the orchid cactus because the plants produce very lovely blooms. The branches are usually flattened and leaf-like and the fruits that follow the flowers are quite delicious. In its natural habitat, this sub-tribe likes to grow on other plants, and because of this, when grown in pots, epiphyllums prefer partial shade in the summer and full sun in the winter.

E. pittieri – produces slightly hanging branches bearing green and white flowers, which may be 7 in. long and 6 in. across, opening at night and hyacinth-scented. Numerous hybrids are available today, such as Toledo, Jean Dupois, and Bahia. These are all worth growing.

GYMNOCALYCIUM In this genus the plants may be found in a clump or solitary. They have clearly visible ribs which are generally notched with warts. The spines are strong and may be awl-shaped. It is a free-flowering genus as a rule, the fruits being usually reddish.

G. denudatum – known as the Spider Cactus because of the particular formation of its spines. Grows to a height of 4 in. and has a globular stem of a deep green colour. The flowers are 3 in. long, green outside and usually pale rose within.

G. saglione – recommended because it is very quick-growing. The stems are often a foot across and the spines are reddish or blackish. It is not, however, free-flowering.

HARRISIA This is a genus of weakly-growing plants: the stems may start by being erect, and then become prostrate. The flowers are seen at night and are borne at the tips of the growths. The spines are like needles.

H. guelichii – another fast-growing plant. The stems are thin and straggling, usually 2 ft. long and 2 in. thick. The flowers may be 10 in. long, green outside and white inside.

H. tortuosa – the easiest of the harrisias to grow, and will put up with very indifferent treatment. It grows erect at first, but later arches and sprawls. The stems may be 2 ft. long and 1½ in. thick. The flowers are brownish green outside and white within.

LEMAIREOCEREUS A very similar genus to *Cereus*, the plants being columnar or branching. The medium-sized flowers appear in the day and are bell-shaped, and the fruits are often edible.

L. marginatus – called the Organ-pipe Cactus, because it branches from the base. Fast-growing and the stems may eventually reach 6 in. across. Its greatest height is probably 25 ft. though you seldom see this in Great Britain. The flowers are red outside and greenish white inside.

L. stellatus – a cactus whose fruits are much liked by the Mexicans. The stem may be 4 in. thick and is covered with white wool. The flowers are pinkish red and last several days.

MAMMILLARIA A very large genus containing over 200 species, getting its name from the prominent warts which it bears and which are often hairy or woolly. The flowers appear in the day and grow in the axils of the old warts. Of the many Mammillarias it is possible to grow, I would recommend the following:

M. elongata – fast-growing, but only reaches a height of 4 in. Its stems are bright green and cylindrical and the flowers are whitish or yellowish.

M. magnimamma – a cluster-forming cactus with globular white woolly stems. When wounded it exudes a milky juice. The stems are 6 in. across and the flowers are green-coloured.

M. wildii – another cluster-forming variety which only grows to a height of 6 in. and is very popular with flat dwellers because it likes centrally-heated rooms. The stems are cylindrical and woolly and the flowers are whitish inside and have a red band on the outside.

MELOCACTUS The stems are globular or barrel-shaped, usually with bristles or hairs at the tips, and small pinkish flowers opening in the afternoon.
M. communis, the Turk's Head Cactus – bluish-green with a short cylindrical stem which may grow to a height of 3 ft. The spines are pink and turn yellow later, and the flowers are pinkish, tipped with red.

OREOCEREUS These are usually called the mountain cacti. The plants branch from their base, forming clusters of stems which are never large, but often very hairy. The flowers open in the daytime.
O. trollii, the Old Man of the Andes – a cluster-forming type growing to a height of 3 ft, and covered with a dense white wool. The flowers are brownish-red outside, and pure red inside.

RHIPSALIDOPSIS In their natural state the plants of this genus are found in the crevices in rocks or on trees. They are usually much branched and often have aerial roots.
R. rosea – resembles the Christmas Cactus and is simple to grow. It bears hanging, much-branched stems, and pinkish-white flowers are produced from the tops of the joints.
R. cassutha, the Mistletoe Cactus – a hanging plant bearing slender branches and greenish white flowers, followed by white round fruits.

SELENICEREUS An easy-to-grow genus, with beautiful flowers opening at night-time. If there is a complaint, it is that the plants are apt to grow too large and that they do not flower until they are pretty old.
S. grandiflorus, Queen of the Night – stems usually 1 in. in diameter and partly covered with yellowish wool. The plants may grow to a height of 15 ft, and the greyish green stems produce beautiful 7 in. long flowers, white inside and salmon pink outside. They open in the evening and die before morning. They are highly scented.

ZYGOCACTUS There is only one species in this genus, which in its natural habitat grows on the stems of trees.
Z. truncatus, the true Christmas Cactus – produces many hanging branches which bear, at the tips, deep red fuchsia-like flowers from late October until mid-January.
For those who want further information about the hundreds of species of cacti that may be grown, reference may be made to my *Cacti as House Plants*.

Above, growing tomatoes and potatoes in whale-hide pots. *Below,* firming a strawberry plant, transferred from a small to a large pot. Note position of plant to one side of pot, not in the centre.

TOMATOES. *Top*, pinching out side-shoot. *Middle*, stopping a plant by pinching out the growing-point. *Below*, growing in Tom-Bags.

Succulents

The following is a selection of ones which are fairly easy to grow.

AGAVE Often known as the century plant. Some species grow very large. The plants can be put out into the garden in the summer.
A. parrasana – a good species to grow because it is not too tall.

ALOE This is often grown for its attractive foliage and most varieties can go outside during the summer.
A. variegata – known as the Partridge-breasted Aloe, because it bears triangular dark green leaves with white markings.
A. distans – also easy to grow. It has short stems and dense red flower-heads.

FIG. 25 *Aloe variegata*

ARGYRODERMA This is the name given to the stemless plants which look like split-open pebbles. No water should be given between December and April.
A. octophyllum – yellow flowers.

BRYOPHYLLUM This is a small shrubby plant with succulent leaves and red or yellow flowers produced in winter at the ends of the stems.

B. crenatum – grows 18 in. tall and has round grey-green leaves with scalloped red margins.

CONOPHYTUM Small cone-shaped succulents which generally produce fragrant flowers. The plants must be kept absolutely dry in May and June.

C. minutum – a tiny grey-green plant with purple flowers.

C. truncatellum – a spotted pale green species with yellow flowers.

CRASSULA A small succulent shrub whose leaves are usually in rosettes at the base of the stem. The plants like to be watered freely during the summer.

C. argentea – egg-shaped leaves and pink flowers.

C. corallina – small shining leaves with yellow flowers and trailing stems.

C. schmidtii – low-stemmed with thin leaves and rosettes, and deep red flowers.

ECHEVERIA This has succulent leaves varying in colour and shape, but always in rosettes.

E. retusa hybrids – one of my favourites, with bluish-green red-edged leaves in large rosettes, and crimson goblet-shaped flowers in the winter.

E. setosa – rosettes of greenish leaves covered with white hairs. The flowers are red with yellow tips.

EUPHORBIA These are usually called spurges, some plants of which are taken for cacti. The flowers are insignificant.

E. globosa – globular, but bears warted spineless branches.

E. grandidens – tree-like, producing small yellowish flowers.

E. pulvinata – forms dwarf spiny cushions with globular stems.

E. splendens, the Crown of Thorns – so called because it is supposed to be the plant from which the crown of thorns was made for Our Lord. It bears very long spines on stout stems, often 3 ft. high, and produces bright scarlet little flowers.

HAWORTHIA Look like aloes, but are smaller. They need to be re-potted every two or three years.

H. fasciata – stemless but has rosettes of thick leaves.

H. rigida – 4 in. tall and has leaves arranged in spirals.

H. hystrix – purplish or greenish stems, porcupine-like spines, and yellow or red flowers.

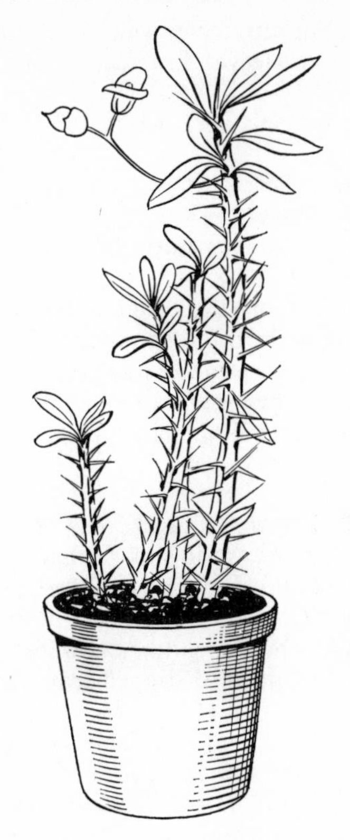

FIG. 26 *Euphorbia,* the Crown of Thorns, with blood red flowers

KALANCHOE These are small shrub-like plants with succulent leaves. They do particularly well in Levington compost.
K. blossfeldiana – dark green leaves with red margins and small scarlet flowers. It can be raised from seed grown in the spring.
K. tomentosa – erect stems with rosette-like leaves at the top.

LITHOPS These are the true pebble plants with which I have had great success. They need only a little water from May to December. The flowers may be white or yellow, and are often very much bigger than the 'pebble' itself.
L. fulleri – very small, greyish with brown markings.
L. marthae – bigger than *L. fulleri,* often 1 in. high with a rounded greyish top.
L. olivacea – dark olive colour, with a deep central groove.

PLEIOSPILOS An attractive plant. Useful in the home.
P. nelii – thick stone-like leaves and yellow flowers 3 in. across.

PACHYPHYTUM Look very much like echeverias, with which they hybridise very easily.
P. bractosum, the Silver Bract – it grows 12 in. high, has greyish-white leaves and bright red flowers.
P. nelii – thick stone-like leaves and yellow flowers 3 in. across.

SEDUM Trouble-free on the whole, and easy to grow – many species are very suitable for the rock garden. For the greenhouse, try:
S. dendroideum – round leaves and large yellow flower-heads.
S. stahlii – blunt egg-shaped leaves and yellow flowers.
S. pachyphyllum – an erect branching plant with thick leaves tipped with red, and yellow flowers.

SEMPERVIVUM Known as houseleeks or sometimes St Patrick's cabbages. They are often grown in sink gardens as they are quite hardy, but they also can be included in a greenhouse succulent collection.
S. arachnoideum – the Cobweb Cactus – so named because the leaves are covered with hairs which look like cobwebs.
S. tectorum – rosettes of purplish-tipped leaves and dark red flowers.

SENECIO Good house plants, variable in shape and form. Water should only be given once a month in winter.
S. galpinii (*kleinia*) – erect, having pointed leaves and bright orange flowers.
S. stapeliiformis – erect with awl-shaped stems and bright red flowers.

TRICHODIADEMA A small shrubby plant with cylindrical leaves which are barbed at the end. The species are very easy to grow.
T. bulbosum – grows 8 in. high and has thick leaves crowned with white hairs.
T. stellatum – cylindrical green leaves, also crowned with white hair.
T. stelligerum – erect cylindrical leaves crowned with brownish bristles.

PART FOUR

Food Production in the Greenhouse

CHAPTER 19

Vegetables and Salads

SOME people like to use greenhouses for growing vegetables and salads, with the idea of trying to maintain a supply all the year round. Within reason almost all vegetables can be grown under glass, providing they are given the right conditions, but generally speaking the amateur gardener concentrates on growing in his greenhouse the types of vegetable that do not do so well out of doors, as well as growing various salads and vegetables for the winter and early spring when they are particularly expensive in the shops.

The vegetables and salads that are normally grown in a greenhouse are tomatoes, cucumbers, lettuces, French beans, climbing French beans, marrows and squashes, radishes, mustard and cress, cauliflowers and spinach. In addition, large numbers of vegetable plants can be raised in the greenhouse for planting out of doors, e.g. onions, celery, leeks, New Zealand spinach, maize, and even early peas and French beans, plus outside tomatoes, cucumbers, aubergines and capsicums. (Full details of growing plants for outdoor culture will be found in my book *The Complete Vegetable Grower.*)

TOMATOES

Seed sowing. Seeds can be sown any time from mid-November to late May. The first sowing results in an early crop which is expensive to grow while the second produces a crop which, though late, needs very little heat. Most gardeners aim to sow about 20th January to obtain a crop lasting about from 20th June to 20th September. This allows the greenhouse to be used for bulbs in the early spring and chrysanthemums in the late autumn and winter.

It is convenient to sow the seed in the standard-size seed trays, and later pot the seedlings into 3 in. pots. The boxes should be filled with Levington or Alexpeat seed compost which is then generally pressed down with a presser board so that it is level and within half an inch of the top. Two methods of sowing are recommended.

1. Sow about 200 seeds per tray and prick-off the seedlings into 3 in. pots as soon as the strap-shaped seed leaves are fully expanded horizontally. This stage should be reached in about eight days if the temperature is kept at 70°F by day, and 65°F by night.

2. Sow 35 seeds per tray, spaced out at 2 in. intervals, i.e. 7 rows of 5 seeds per tray. The resultant seedlings remain in the trays till they are about 1½ in. high, i.e. when two pairs of true leaves appear, which should be about 28 days after sowing. At this stage they are potted off into 3 in. pots. The seed should be covered with ¼ in. thickness of the seed compost or, if desired, you can press each seed into the soil with a pencil before covering with the further compost or sand.

Germination is slightly slower in the second method than in the first, and therefore such thin sowings should be made a week earlier than thick sowings if you want plants fit to put out at approximately the same time. On the other hand, the seedlings from the thick-sowing method have to be potted up earlier in life and then occupy more space for a longer period. There is little doubt that better plants arise from the thin-sowing method.

The boxes should be covered with a sheet of glass and newspaper, which are taken off each morning and replaced after the condensed moisture has been removed from the glass. Germination should take place in about 3-4 days and then the glass and paper should be removed altogether and the seedlings placed as near the greenhouse glass roof as possible.

If you are planting in the soil of the greenhouse at ground level, the ideal is a friable, well-drained but firm, sandy loam. Most gardeners, however, have to put up with the earth as it is. Sedge peat or well-rotted compost can be dug in very shallowly at 3 bucketfuls to the square yard, and when planting on the flat, lightly fork in 6 oz. of fish manure with a 10% potash content to the square yard. Ensure that the soil is thoroughly moist to full rooting depth before planting.

Planting out. The subsequent treatment may be divided into two groups: planting in the border and planting in pots. When turning out the plants from the pots the ball of soil should be carefully kept intact and only the crock removed.

In the border, most varieties can be planted in rows alternately 1 ft. 9 in. and 2 ft. 3 in. apart, the plants being 15 in. apart in the rows. From such planting, with proper treatment, an average crop of 5 or 6 lb. per plant may be obtained. Water the plants well immediately before and after planting, but be careful not to water actually on the stem.

Trimming. All side-shoots should be pinched or cut out immediately they appear. Dense foliage may be thinned in order to allow for movement of air and action of sunlight, but ruthless cutting away of the leaves should, however, be avoided. Any yellowing leaves which appear should be cut off and put on the compost heap.

Staking. All tomato plants in a border need support, either with bamboo canes or with string. In the case of the string method, overhead wires are strung tightly at a height of 6 ft. and the string is tied to this. The other end of the string is then tied to the base of the plant or to a short bamboo cane driven in near the base, and as the plant grows and needs support the string is just twisted round it.

Watering. It is very difficult to lay down any hard and fast rules for watering. The gardener will have to be guided by the look of the plants as well as the atmospheric conditions. If the leaves seem to be flagging, it is a sure sign that water is needed, and water should be given liberally since light sprinklings do more harm than good. Generally speaking, however, a good soak once a week during the summer months will suffice.

Temperature and ventilation. The temperature in a tomato house should never be above 65–74°F in the daytime, or below 50°F at night-time.

Give as much ventilation as possible during the summer, particularly at the 'ends' of the house. Keep the atmosphere moving all the time but avoid draughts. This is done by using a little heat plus some ventilation. Syringing plain water over the plants helps to ensure that the flowers set properly. Some shading at this time will do good also.

Feeding. One of the greatest faults a novice can make is the constant application of fertilisers during the growing season. Over-manuring results in the dropping of flowers, makes the flavour of the fruit very watery and insipid, and the plant more liable to disease, particularly mildew. Top-dressings of damped sedge may be given as a mulch 1 in. deep, and fish manure with a 10% potash content may be applied in three dressings, using about 3 oz. to the square yard each time. Do this at monthly intervals starting, say, at the beginning of June.

Watering with Marinure (Tomato Special) is an excellent method of feeding tomatoes in the summer. Follow the instructions on the bottle.

In pots. Tomatoes intended for this treatment should be grown for about three weeks in 3 in. pots and then transferred to 6 in. pots for a similar period. They will then be ready for transference to their final 10 in. pots, which may be of clay or whalehide, containing a Levington or Alexpeat compost. In transferring to larger pots it is important to disturb the roots system as little as possible – the ball of soil from the smaller pot should be kept moist and intact and a hole of sufficient size made with a trowel in the compost in the larger pot. The 10 in. pots should be filled with compost to within about 4 in. of the top so as to allow a top-dressing to be added later. The plants should be firmed in and a little bone-meal may be added at 1 oz. to the square yard.

Top-dressing. When several clusters of fruit have been set a top-dressing will be necessary. This should consist of 2–3 in. of Alexpeat or Levington compost No. 2. Six weeks afterwards another such dressing may be needed, and after each operation a good watering will be necessary to settle the compost.

Water and food. Pot plants always require more watering and feeding than plants in the border, and it will usually be necessary to water every day, adding Marinure to the water twice a week. In general, however, the rules for border cultivation apply equally to pot plants.

The Ring Culture Method. The pots used for ring culture of tomatoes, whether they are of whalehide or clay, are stood on what is called 'aggregate': unburnt cinder, clinker, Lytag, or even coarse sand. This is to allow roots on the plant to 'collect' the water it needs from the aggregate area and the fibrous roots in the pots to gather the foods required from the soil.

Pot up the tomatoes into large 10 in. pots of Levington or Alexpeat compost and stand them on the aggregate, which should be 4 in. deep. This may be done from early April onwards. In the early stages water the pots and keep the aggregate as dry as you can. Give frequent small doses of water rather than attempting to flood the pots.

When the roots of the plants are well established in the pots the aggregate should be drenched with water. The 'moisture' roots will then penetrate through the hole or holes at the base of the pots in search of water. This 'pulling up' of the moisture required from below does keep the compost in the pots adequately supplied with water, and especially so if the plants are syringed over every day or every two days or so, if the weather is dull.

The exception to this rule, perhaps, is in mid-summer during a really droughty period, when I usually have to give each pot a pint of water twice a week. This keeps the compost just moist. Feeding with Marinure is usually done twice a week, and a pint of the diluted feed per plant is usually all that is necessary – plus any normal watering to keep the leaves turgid. The feeding starts after the first truss has set its flowers. Sometimes this special feeding makes the compost in the pots rather acid and the answer in this case is to use 1 large tablespoonful of carbonate of lime at every fourth feed. If sprinkled evenly over the top of the compost it will be washed in when the liquid feed is

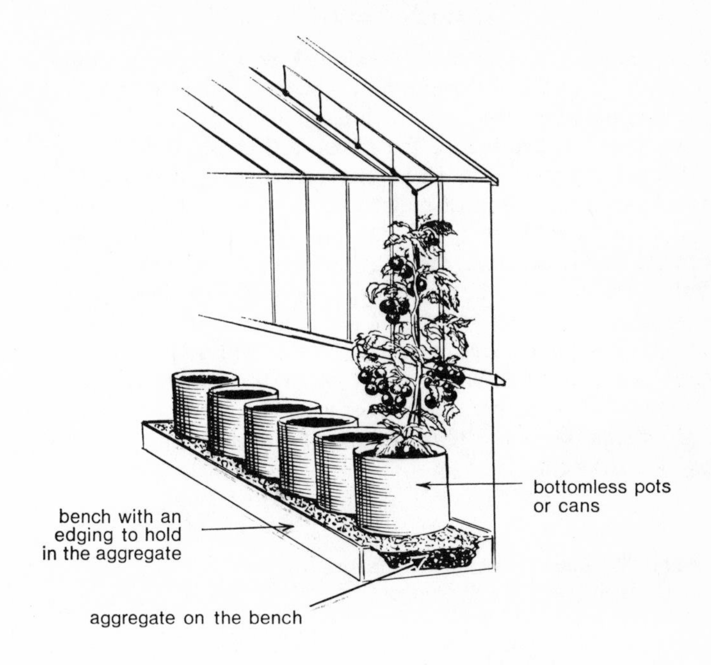

FIG. 27 Growing tomatoes by the ring culture method

applied. If planting is done, say, on 15th April, then the first feeding will probably be about the 20th May, and the first fruits will be ready about 10th July.

The dis-shooting or defoliating of plants grown by this method is similar to that described for plants grown in the border or in pots.

Varieties

Eurocross A or B – immune to the leaf-mould disease
Supercross – Immune to leaf-mould, non-greenback
Golden Sunrise – a yellow variety

FRENCH BEANS One can only grow French beans under glass, as scarlet runners and broad beans find difficulty in setting their 'fruits' under these conditions. Climbing French beans are, however, the ideal substitute for runner beans.

Dwarf – grown in pots

General cultivation. Use 10 in. pots filled with Levington compost No. 2 to within 3 or 4 in. of the top. Sow 8 beans 1½ in deep around the edge of the pot, and after germination, thin the plants down so they are evenly spaced. Don't water before the plants are through the ground – just see that the compost is moist before seed sowing. When the plants are through, keep the compost moist.

When the plants are 6 in. high add some more compost so as to cover the roots and raise the level half an inch. Do this again at fortnightly intervals until the level of the compost is within 1 in. of the top of the pot.

Syringe the plants over every day, as this will help to ensure that the flowers set properly. When the first pods form start feeding with Marinure and continue once a week after this. Support the plants with twiggy sticks, keep the temperature of the house at 55°F and never allow any draughts.

Varieties

The Wonder – particularly delicious
Masterpiece – a heavy cropper

Climbing – grown in the border

General Cultivation. The seeds should be sown in 3 in. pots filled with Levington compost – one seed per pot. The pots are then covered with a sheet of brown paper to keep out the light. Stand the pots on the staging of the greenhouse so that they can get some bottom heat.

Remove the paper when the seeds have germinated, and give the pots a good watering. Wait until the plants are 2 in. high and then plant them out in the soil in the border of the greenhouse where they are to grow. Allow 28 in. between the rows and 9 in. between the plants in the rows. Provide string or bamboo up which the plants can climb. To get very early out-of-season crops the planting should be done about 15th January – though those who live north, say, of Birmingham will seldom be able to plant before mid-February. The temperature in the house at this time should be 65°F.

Make holes with a trowel large enough to take the pots, and put the plants in these holes for two days. Then, when the temperature of the compost has fallen to the same temperature as the soil, knock the plants out of their pots and plant firmly. Then give a light watering.

When the plants are growing well syringe them over at least once a day, and if the sun is very hot, twice a day. When the plants reach the top of the strings or bamboos either stop them by pinching out the growing points or train them on to the top of the next bamboo so as to form a series of arches. Feed with fish manure twice – once in March, and again early in May giving 3 oz. to the square yard. Keep down red spiders at all costs by spraying with liquid Derris from time to time, paying particular attention to the underside of the leaves.

You should be able to get a good crop of French beans by the end of June, removing the plants in time to make room for a late crop of tomatoes.

Varieties

Vertch's Climbing
Guernsey Runner
Early Blue Lake

CAULIFLOWERS Very early cauliflowers are more delicious than winter broccoli, which merely look like cauliflowers.

General cultivation. Sow the seed thinly in a drill half an inch deep in a frame, about 10th September. Rake lightly to cover the seed and firm over the top with the rake head. Give a light watering, put the frame light over the frame and then cover the glass with a hessian mat. When the seeds have germinated remove the hessian and start ventilation until the lights can be removed altogether, say three weeks later.

Early in October pot up the seedlings into 3 in. plastic pots filled with Levington compost and put on the staging of the greenhouse at a temperature of about 45°F.

In January prepare the greenhouse border by forking lightly and adding medium-grade sedge peat at 2 large bucketfuls to the square yard. If the soil is at all dry, flood the border well, first adding carbonate of lime at 7 oz. to the square yard. Leave for a week and then apply a fish manure with a 10% potash content at 5 oz. to the square yard and rake this in.

Make holes with a trowel 2 ft. apart between the rows, and 18 in. apart in the rows. Knock out the ball of soil from the pots, and place firmly in the hole. Water in. The temperature should now be 50°F.

Ventilate whenever possible. Syringe the plants over on bright sunny days, and water well once a week as the plants near maturity. Harvest in April.

Varieties

Romax Extra Early
All-the-Year-Round
Early Snowball
Alpha
Novo

CUCUMBERS The culture of cucumbers (*Cucumus sativa*) is one of the oldest arts, for this vegetable has been popular in China and Egypt for thousands of years.

Seed sowing. Seeds can be sown at any time from early January onwards in 3 in. pots, one seed being placed point downwards in the centre of each pot. The seed compost used should be the Alexpeat or Levington. The bottom of the pot should be well crocked and covered for 1 in. with a rough sedge peat. After sowing, the seeds should be covered with half an inch of the same compost and then watered. The pots should be placed on the greenhouse staging where some bottom heat can be assured. The temperature of the house at this stage should never be allowed to fall below 70°F at night-time.

Potting up. In a fortnight or so the plants should be ready to be transferred to their permanent beds, but if they are not quite ready pot on temporarily into 5 in. pots of No. 1 Alexpeat or Levington compost at the same temperature as the house. Cucumbers should not be potted firmly and each plant should be inserted right up to the seed leaf. A good soaking is again necessary but watering afterwards need only be done when the soil appears dry. Good root development cannot take place in continually wet soil.

Staking. After a week in the pots the plants will need support, and an 18 in. piece of fine bamboo should be inserted into the compost and the cucumber stem tied loosely to it to allow room for development.

The bed. When the plants are 6–9 in. high, they may be transferred to their permanent beds. It is very important to ensure good drainage all the time, and water must be able to get away quickly. Coarse sedge peat or old straw may be used to a depth of 12–18 in. in the borders, but both will need a good soaking a few days before the mounds are put over them. These mounds are made of Alexpeat or Levington compost and should be about 9 in. wide, 4 or 5 in. high and 20 in. apart. After preparing the bed, the greenhouse should be heated to a temperature of 70°F about four or five days before planting. Thus the beds are

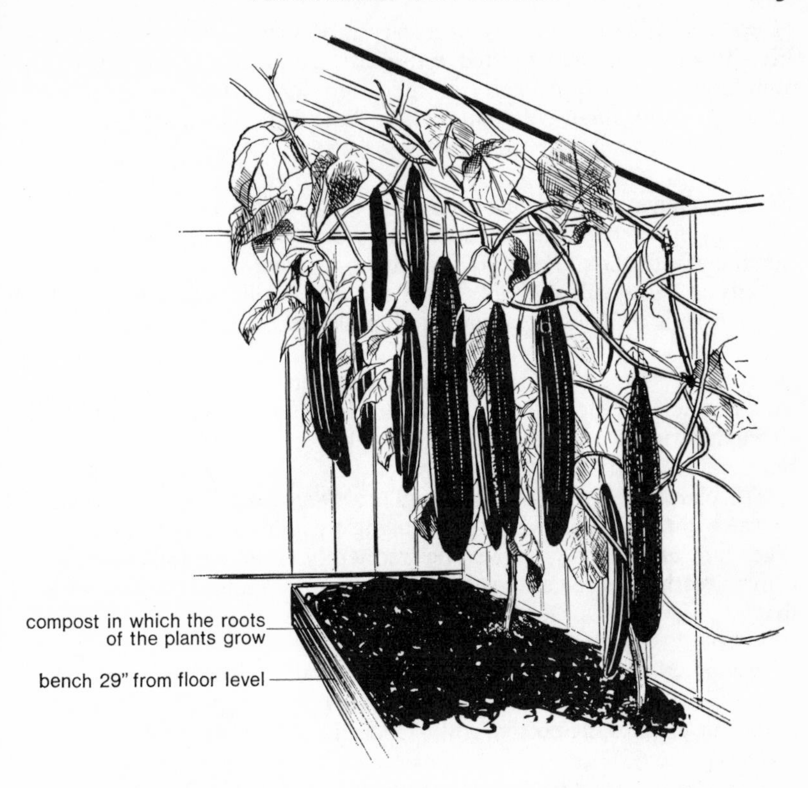

FIG. 28 Training cucumbers up the side of a greenhouse

warmed and ready to accept the roots of the cucumber – the greatest care should be taken that the plants are not chilled when they are moved.

One pot should be planted in the middle of each little mound and should remain there for 24 hours before the ball of soil is knocked out and planted firmly in the hole the pot occupied. The top of the ball of soil should be 2 in. below the surface of the bed. Watering may then be carried out, a gallon being given to every 3 mounds. A bamboo must now be provided for the purpose of training the new growth to the wires or trellis, unless the first cane proves long enough.

Temperature and ventilation. A high temperature should be maintained at all times. At night it should be 70°F, and in the daytime it may ris to 90°F. Ventilation is rarely necessary in the early part of the plant's life and air should be admitted with great caution. Ventilators, if used, should be closed quite early in the afternoon to trap the sun heat. Towards July slight ventilation may be given, particularly on the leeward side, to change the air in the house.

Shading. When the sun gets bright scorching of the foliage may occur. Shading must therefore be provided by spraying whitewash over the outside of the glass. Towards July this may have to be repeated more heavily as rains will probably wash the first application off. A better alternative is to have the automatic blinds described on p. 40.

Watering. As much harm is done to cucumbers by over-watering as by under-watering. Moderate waterings are of little use; soak the bed well through twice a week. If possible the water should be at the same temperature as the house.

The plants should be syringed regularly twice a day, the usual amount of water in the mid-growing season being 1 gallon to every three plants. The walls and paths must also be frequently syringed over, especially in hot weather, as the atmosphere in a cucumber house must never be dry.

Trimming. No laterals must be left on the young plants below the first wire. The growing point of the plant may be 'stopped' at the fifth wire; this enables the bottom fruits to swell quicker, and so gives earlier fruiting.

There are a number of rules which should be followed:

1. Do not allow a cucumber to grow on the main stem.
2. Do not allow the side-growth to grow further than the second leaf before stopping.
3. Do not take a cucumber and a growth at the same joint; the ideal should be first a cucumber, second a growth where it is stopped, and so on.
4. A good idea is to stop the main laterals at two joints and the sub-laterals at one joint.
5. Aim at having two fruit-bearing joints on every lateral, and not more than three breaks.
6. Rub out all the male flowers as these affect the flavour and shape of the cucumbers. The female flowers have a very tiny cucumber at the base. The male flowers are borne on a thin stem.
7. Remove tendrils, dead fruitlets and yellow leaves.

Training. All growths must be tied in to the wires loosely or growth will be impeded. All young fruit should be kept clear of ties and wires and should hang cleanly downwards.

ARROWS. *Top*, sowing seeds in peat pots. *Middle*, pinching out the second, unwanted seedlings. *Below*, watering through a flower-pot sunk in the soil.

Above, the 'Turk's cap' form of gourd. *Below,* pinching off a second, unwanted flower truss from a grape vine.

Top-dressing. Top-dressings and mulchings are essential and should be given whenever required. The first top-dressing may be given 10 days or so after the planting date. It is usual to top-dress again as soon as the white root fibres are seen coming through the surface of the soil. Only a little compost should be put on every time. The compost used for top-dressing should be the No. 3 Alexpeat or Levington.

Varieties

Butcher's Disease Resisting
Telegraph Selected
Green Spot
Princess and Simex are useful because they produce all female flowers after the first few weeks.

LETTUCES Some experts try to have lettuces available almost every month of the year by growing them in a cool house during the winter. A good lettuce house is one where the plants can be grown in the natural soil at ground level, and where there is plenty of light to ground level and sufficient air and height of house. One can almost say: the bigger the house, the better.

It is most important to grow the correct varieties, since the ordinary outdoor kinds just will not do. The two main cropping periods should be autumn and early spring.

It is convenient to use the lettuce as a 'catch' or follow-on crop to tomatoes, i.e. the lettuces are grown before the tomato plants are set in position in the spring, and after they are pulled out, maybe in late September. On occasions it is best to plant out the young tomatoes while the lettuces are still growing, and to see to it that the lettuce planting is done so as to leave room for properly spaced tomato plants.

There are two main plantings, the first between mid-September and mid-October, yielding good-hearted specimens in December and January, and the second in December and January, producing fairly good lettuces in February and March.

Seed sowing. For the winter crop the seed may be sown between 20th August and 15th September in seed trays filled with Levington or Alexpeat compost, 200 seeds per standard seed box. The compost is levelled in the box so that it is within a quarter of an inch of the top, the seed is then sown thinly, a little more of the same compost sifted over them and pressed down. A light watering through a fine rose is then given.

Cover the boxes with a sheet of glass and a piece of newspaper, and place them on the benching of the greenhouse at a temperature of

60°F. Germination should take place in four days, and then the glass and paper are removed. When the seedlings are ¼ in. high, prick them out into further seed trays filled with a similar compost, at 54 seedlings to a tray. Put the seed trays back on the staging of the house at a temperature of 55°F and water lightly.

For the spring crop sow the seeds about mid-October and treat in a similar way.

General cultivation. The lettuces should be planted out 9 in. apart in the soil of the greenhouse, early in October for the December and January cutting, and about mid-December for the spring cutting. In both cases rake the top soil level and add fine sedge peat at 3 or 4 bucketfuls to the square yard. Plant firmly but shallowly, and after planting water well through a fine rose, holding the can low down so that the water goes on the soil and not on the leaves of the plants. Water regularly in this way so as to keep the leaves turgid. If they are allowed to droop on the soil, botrytis (a fungus disease) will get in through the breathing pores of the leaves. Give a specially generous watering 3 weeks before you expect to harvest. This reduces marginal leaf scorch.

Remove all decaying and dying leaves the moment they are seen, and put them on the compost heap. The temperature of the house should never exceed 55°F at night, though it may be allowed to rise in the daytime to 60–65°F if this rise is due to the sun. Ventilate on all favourable occasions, and put on as little heat as possible.

Hoe lightly between the rows three times during the growth period to keep down weeds.

Varieties. Special ones for greenhouse growing are:

Hilde
Cheshunt Early Giant
Kordaat
Kwiek
Maziola – sow August to cut in November
sow October to cut in April
Deciss – sow September to cut in December-January
Vitesse – sow mid-September to cut in February-March

MARROWS AND SQUASHES Regard these as luxuries produced at the 'wrong' time of the year. Grow one of the trailing varieties, which can climb up any purlin posts there may be in the house or, if necessary, poles or strong wires provided for the purpose.

General cultivation. Sow the seeds singly in 3 in. plastic pots filled with Alexpeat compost. Give a watering, put the pots on the staging of the greenhouse and cover with a sheet of glass and a piece of newspaper. When the seed leaves appear, remove the glass and paper and see that the plants get all the sunshine they can.

Dig a hole near the base of each post or wire up which the plants are to climb, place in the hole a 6 in. pot filled with sedge peat, plant the marrow or squash firmly in it and water well.

As the plant grows upwards, tie the stem loosely to the support, and when it reaches the top of the house pinch out the growing point. This will result in the development of side-growth. Look for the female flowers – those with a tiny incipient marrow situated between the stalk and the yellow flower – and with a camel's-hair brush, take some of the pollen from the centre of the male flower and transfer it to the centre of the female one.

When the marrow grows large it must be supported or else it will pull the plant down by its weight. Just loop a piece of string round the stalk of the fruit and tie it up to the post or wire. If there seems to be too much foliage cut some of it off. Give the plants plenty of water.

Cut the marrow or squashes when they are 12 in. long.

Varieties

MARROWS
Long Green
Long White Trailing
Long Green Trailing
SQUASHES
Delicious Golden
Summer Crookneck

MUSTARD AND CRESS No plants mature quicker than mustard and cress. If the temperature of the greenhouse can be kept at 60°F, it is possible to cut this crop within 8 days of sowing the seed. (Rape is often preferred to mustard. It tastes more delicious, and grows quicker and more abundantly.)

General cultivation. The seed can be sown thickly in the soil of the greenhouse after it has been raked level, and fine sedge peat added at the rate of 3 bucketfuls to the square yard. It can also be sown in sterilised soil on the benches, in Levington compost in boxes, or even on damp sacking laid on the soil or the benches.

The cress seed must be sown two days before the mustard, and most people agree that the proportions should be $\frac{1}{3}$ cress and $\frac{2}{3}$ mustard. The seed is just pressed into the soil or compost with a wooden presser or, if sown on the sacking, just watered well.

Water regularly so as to keep the compost, soil or sacking just moist. It pays to try and keep the mustard and cress in the dark for three or four days after sowing, for example, by covering the sown seed with brown paper.

POTATOES This is another luxury crop grown by those who love very early new potatoes.

General cultivation. Choose potatoes the size of a hen's egg when the early outdoor varieties are lifted in the summer. The tubers must come from virus-free plants and are then placed in seed trays and kept in an airy shed free from frost.

Before planting, the seed tubers are placed 'rose end' upwards in seed boxes and put in a greenhouse at a temperature of 60°F, so that the tubers may sprout. By the middle of January each sprouted potato should be planted in the centre of a 12 in. pot which has been well crocked and then filled with Levington compost. The pots are then placed in a greenhouse at a temperature of 45°F. When the plants are 4 in. high the temperature may be raised to 50°F, and when they are 8 in. high liquid manure with a high potash content like Tomato Special can be given every five days until harvesting time. Water as necessary to keep the compost moist. The rose-end is the end with the greatest number of eyes.

If the planting is done about 10th January the new potatoes should be ready from the end of April onwards, but do not expect a heavy crop or large tubers.

Varieties

Ninetyfold
Arran Pilot
Arran Comet
Pentland Beauty

RADISHES You must never attempt to force radishes. They must be grown as naturally as possible, and it is essential to grow the right varieties. Some amateurs get three good crops of radishes between the middle of September and the end of March.

General cultivation. Sow the seed in mid-September in the greenhouse bed after raking the soil level and adding fine sedge peat at 3 or 4 bucketfuls to the square yard. This organic content of the soil is important. Apply wood ashes (if available) at ½ lb. to the square yard, and bone-meal at half this rate. Rake in well. Now flood the border thoroughly and when, in a day or two, the soil has dried sufficiently sow the seed evenly at 1 oz. to 3 square yards; gently press it into the soil or rake lightly. The temperature of the house should be 50°F.

Give a light watering the moment the seedlings are through and keep the greenhouse well ventilated. The first sowing should be ready for pulling in October. The soil is then raked down level again and a second sowing made. When this second crop is harvested the third sowing is made.

Varieties

French Breakfast Short Top
Earliest Scarlet Short Top
French Breakfast Forcing
Woods Early Frame

SPINACH

General cultivation. Lightly fork over the soil in the greenhouse border and add medium-grade sedge peat at 3 two-gallon bucketfuls to the square yard. This peat should be well soaked in water before use if the weather is dry. Make the border moderately firm by treading and follow this by a very light raking to make sure both that the surface is level and that the surface particles of soil are finer than grains of wheat.

Sow the seed in rows 10 in. apart, and in drills ½ in. deep, then rake lightly to cover the seed. Water thoroughly through a fine rose going over the rows again and again so as to soak the soil but not disturb the seed. Sow in mid-September for harvesting in the winter, and in January for cutting in the spring.

When the seedlings are 2 in. high thin them out to 6 in. apart. After this, water the rows roughly once a week, but never splash soil on to the leaves of the plants, and also make certain that the leaves have dried off before the sun sets.

In the case of seed sown in the autumn the house must be heated in the winter to a temperature of 50°F. With the January sowing the temperature can be as low as 40°F.

Varieties

Nobel
Monarch Long Standing
Hurst's 101

CHAPTER 20

Fruits

THERE are a number of fruits which can be grown out of doors but which do better in the greenhouse, as they are indigenous to warmer countries than Great Britain. I refer particularly to vines, peaches and melons. Other fruits, e.g. strawberries, ripen far earlier when grown under glass.

Although some amateurs have had unusual success with growing a vine-rod or a peach, for example, against the wall, in addition to plants such as tomatoes in the summer and chrysanthemums in the winter, the general plan should be to devote one whole greenhouse to vines and another to peaches. Melons, however, can be grown with cucumbers, and strawberries in pots can be forced on the shelving of any greenhouse while other plants are grown on the benches.

More people use their greenhouses for flowering plants and salads such as tomatoes, than for fruits, but there are always a few who feel they want to produce their own bunches of grapes and luscious peaches. Simple instructions are given below for the growing of the most delicious varieties in the simplest possible manner.

FIGS The simplest way of growing figs in the greenhouse is to grow the tree-fan trained up the back wall of a lean-to greenhouse, or even up special strong wire trellis erected perpendicularly in the centre of a normal greenhouse. The alternative is to grow a fig tree in a 12 in. pot and to keep it small by summer pruning.

Planting. Dig a hole 3 ft. long, 18 in. deep and 3 ft. wide. In the bottom of this place broken bricks and ram them well down – they will act as a partial barrier to the roots and help drainage as well. Fill the hole with good soil, plus two 3 in. potfuls of bone-meal, two of wood ashes and one of carbonate of lime.

In November or December plant the fig tree in the middle of the hole, spreading the roots out evenly and carefully. Tread the soil down well so as to firm the earth over the roots. Tie the small branches up fan-wise to the wires with plain string.

General cultivation. The trees start into growth in mid-January, as long as they have some heat, about 60°F, at night-time. The branches should be syringed over in the morning and afternoon each day.

When the side-shoots, or laterals, have produced four leaves, the end of the shoots must be pinched back by an eighth of an inch. Any sub-laterals that are produced on the main laterals are stopped in the same way. When the tree is three or four years of age, cropping may commence. Little fruits which had the appearance of buds the latter end of the previous season will be seen forming on last year's wood. These fruits will start to swell, and to help them, water the border regularly when the fig is growing. Marinure may be given twice a week from early May onwards, and you should be able to harvest the first crop of ripened figs in June.

If the house is kept at 60°F at night-time and plenty of ventilation allowed, a second crop of figs may be expected in August. The fruits this time are borne on the young wood, i.e. the stopped laterals. This time the atmosphere of the greenhouse must be kept moist and the tree must be syringed over frequently until ripening starts. The temperature may be allowed to rise up to 90°F in the daytime, and the door of the greenhouse must be shut about 6 p.m. each day and the ventilators closed, thereby 'bottling up' the daytime heat. Keep the glass perfectly clean as light is essential.

Once the second crop has been picked the house may be given all the ventilation possible. The temperature should be reduced to 40°F, resulting in the leaves falling in the autumn, and the trees can then be exposed to a temperature as low as 35°F.

Pruning. The roots of the fig should always be restricted. Therefore after seven or eight years the spade may be plunged into the soil 3 ft. away from the main stem, and all round. Any roots met with during the operation should be cut off.

Fortunately one cannot completely ruin a fig tree by pruning. In winter remove dead wood and the crossing and rubbing branches, and thin out branches where they appear to be overcrowded.

Varieties. Good ones to grow under glass are:

Brown Turkey – an easy-to-grow early
Negro Largo – a mid-season, thin-skinned variety
Brunswick – a large late

MELONS Melons may be grown in a similar manner to cucumbers, and in the same house.

Seed sowing. Sow two seeds ¼ in. deep in January or February, in 3in. plastic pots filled with Levington or Alexpeat compost. If both should grow, thin down to one seedling per pot. It helps if the compost is warmed before being used. Stand the pots over the hot-water pipes so as to ensure bottom heat, or over the electrically-heated base to a propagating frame. Cover the pots with a sheet of glass and a piece of newspaper. Keep the temperature at 75°F and the seedlings should appear in two or three days.

Potting up. When the plants are 3 in. high they should be grown on shelving near the glass, and are ready for potting on into 5 in. pots of Levington or Alexpeat compost No. 2. It is at this period that the plants need syringing over every day, and the atmosphere of the house should be kept on the moist side.

Planting. Make up beds as advised for cucumbers (see p. 222). Plant the melons, when knocked out of the pots, in mounds of compost 18 in. apart. It pays to provide a 1 ft. long bamboo for each plant and to tie the top of it to the first wire. Other wires 1 ft. apart should also be provided, and the melon trained up and along them.

General cultivation. Never water at the base of a plant as this invariably causes the stem to rot at compost level – the dreaded disease known as collar-rot. To prevent this trouble, it helps if the ball of soil from the pot is planted so that it is only half buried. Some gardeners knock the bottom off the pot so that the specimen can be planted in the pot, thereby protecting the collar, but then one must take great care never to water inside the pot.

Allow the plant to grow upwards, and when the leading shoot reaches the second wire pinch back the growing shoots by a quarter of an inch. Side-shoots will now develop which will be tied to the wires with soft string (fillis). As female flowers, recognisable by tiny melons at the base, are produced on the laterals, transfer pollen from the male flowers to the centre of the female flowers with a camel-hair brush. If bees are allowed to enter the greenhouse hand pollination is unnecessary.

Do not allow more than three melons to develop on any one plant, and aim to have the three fruits swelling at the same time. Keep the ventilators open in the middle of the day so as to allow a free circulation of air.

Syringe the plants over twice a day – but on the whole keep the roots drier than for cucumbers. When the melons are actually flowering only syringe once a day. When the fruits are changing colour and ripening stop watering unless absolutely necessary and give plenty of ventilation. The aim should be to keep the temperature of the greenhouse at about 70°F all through the plants' growth.

As the fibrous roots penetrate to the surface of the mounds in which the plants are growing, give top-dressing of Levington or Alexpeat compost to cover them. As the fruits swell diluted Marinure is given once a week at the rate of 2 gallons per three plants. Support the ripening fruits with nets to prevent them breaking the stems, or tie the stem of the ripening fruit to the wire with raffia to take the weight of the lateral.

Varieties

Blenheim Orange
Hero of Lockinge

PEACHES As nectarines are really a smooth-skinned type of peach, what holds good for the one also applies to the other. In the greenhouse it is usual to grow fan-trained trees against the back wall in lean-to structures, or up against perpendicularly-placed strong wire trellis running down the centre of a span-roof house. The fan-shaped trees may be bought as three- or four-year-olds with a stem about 18 in. tall.

Planting. Dig out a hole about 4 ft. long, 3 ft. wide and 2 ft. deep. Into the bottom put plenty of broken bricks, large stones and coarse rubble, and tread down firmly. Above the stones put a layer of coarse sedge peat 3 or 4 in. thick and then fill the hole up with a compost consisting of 6 parts good garden soil, 1 part rough chalk, ½ part wood ashes and ⅛ part of bone-meal. Firm this down well and follow with a really good watering.

In the centre of this hole and about 18 in. away from the wall plant the tree in November. Spread the roots out shallowly but evenly, and tread them in really firmly. Never plant peaches deeply. In January apply damped medium-grade sedge peat all over the ground around the stem of the tree to a depth of 3 in. and a width of 3 ft. Now shorten the one-year-old peach growths to half their length.

General cultivation. Though all varieties of peaches and nectarines are self-fertile it helps at blossoming time to pollinate the flowers artificially with a camel-hair brush. The temperature of the house at this time should be 50°F. When the little fruits start to swell, the temperature may be increased to 60°F at night, and during the day (owing to sun heat) the temperature may be allowed to rise to 75°F.

Apply dried blood around the trees at 2 oz. to the square yard each February and give a fish manure with a 10% potash content at 3 oz. to the square yard each December. Unless the soil of the greenhouse is known to be alkaline, each year hydrated lime must be given at 4 oz. per square yard after the peaches have been picked.

See that one good flooding is given in January, and apply plenty of water in the spring and summer to prevent the soil drying out. In addition syringe over the fan-trained tree after the fruit has set, stopping when the fruit starts to ripen. After harvesting, start syringing again until all the leaves have fallen. As the original sedge peat mulch may have been pulled into the soil by the worms, a further mulching of sedge peat may be necessary in March the following year. Withhold water gradually from October to January.

Thinning. Just before the fruits attain the size of walnuts, thinning must be carried out so as to leave the peaches about 1 ft. apart. Soon afterwards summer pruning should be done to remove unwanted shoots. The others must be tied to the wires with raffia. The idea is to leave a good shoot at the base of each length of fruiting wood, another young shoot half-way up this length of wood, and a third shoot right at the end. The fruiting wood should then be pruned back in the winter to a point just above the basal bud, and the young basal shoot then becomes next year's fruiting wood.

The winter pruning is aimed at producing good lower branches and not allowing the centre of the fan to be filled in for four or five years. The leaders or end-growths and the resulting growths must be pruned back by about half for the first three or four years, then spaced out 15 in. apart and tied to the wires in position. As the tree gets older the pruning consists of cutting out some of the old wood so as to retain the younger, less strong shoots.

Varieties

PEACHES

Amsden June – ripens mid-July
Duke of York – mid-July
Hale's Early – end July
Kestrel – early August
Rochester – mid-August
Sea Eagle – end September

NECTARINES

John Rivers – mid-July
Early Rivers – end July.
Lord Napier – August
Pineapple – early September
Humboldt – mid-September

STRAWBERRIES The whole point of growing strawberries under glass is to be able to pick the fruits months before those out of doors are ready.

Propagation and planting. Prepare a strip of ground out of doors by forking it over lightly and adding medium-grade sedge peat at 3 bucketfuls to the square yard, plus a fish manure with a 10% potash content at 4 oz. to the square yard. Tread the ground well afterwards, or lightly roll, and plant out recently struck strawberry runners from a virus-free strain in August. The rows should be 2 ft. apart, and the runners 18 in. apart in the rows. This early planting in the open is most important. The following March mulch the soil with a complete carpet of medium-grade sedge peat 1 in. deep. When the blossom trusses appear, pinch them off.

When the runners start to grow out from these young plants, peg them down at the nodes into the sedge peat with wires shaped like hair-pins. If the weather is dry, give a good soaking by means of an artificial rain sprinkler. When the runners are fully rooted in the sedge peat, sever them from the parent plants and pot them up into 6 in. pots containing Levington or Alexpeat compost No. 2. Do not plant in the centre of the pot as is usual, but towards the edge. This will ensure that the fruits as they form hang over the sides of the pots.

Place these potted-up specimens in the shade for a week or ten days, and afterwards stand them on a concrete or ash base in full sun. Syringe the plants night and morning in this position and water regularly. Twice a week add Marinure with a high potash content at the rate of ¼ oz. to a 2½ gallon can of water.

Towards the end of October lay the pots on their sides, and 14 days later plunge them in medium-grade sedge peat in a cold frame.

General cultivation. During the first week of January remove the plants from the frame and stand them upright near the light on shelving in the greenhouse. The temperature should now be 45°F at night-time, and may rise to 55°F in the day. For the first fortnight water lightly and syringe the plants over once a day. After that be a little more liberal and do the syringing twice a day.

Gently move a camel-hair brush inside the flowers to ensure perfect pollination, and increase the temperature to 65°F from, say, 9 a.m. to 4 p.m., and 55°F from 4 p.m. to 9 a.m. In all probability too many fruits will be formed, and not more than fifteen good berries should be left per pot. Those over this number should be cut off with a pair of scissors. The fruiting trusses should now hang over the edge of the pot where they can easily be seen.

Stop syringing the plants over as the strawberries start to colour, but water every day and add liquid manure two or three times a week. The temperature now can be as high as 70°F during the day.

Varieties

Royal Sovereign – S.V. 55 – this virus-free strain is probably the best for greenhouse work

Cambridge Rival 632 – produces large early fruits of good flavour and is a good forcer

VINES Many people are fascinated by the idea of growing grapes under glass but there are problems which are often not foreseen. Birds, for instance, can be a nuisance, and will enter the house in large numbers unless small-meshed wire netting is placed over the openings of the ventilators. The thinning out of the berries can be quite laborious, as can the correct summer pruning of the laterals.

Planting. A trench should be made as advised for peaches, and the brickbats put in together with the thick layer of coarse sedge peat. The remainder of the hole should then be filled in with a compost consisting of 6 parts good garden soil, 3 parts fully composted vegetable waste, $\frac{1}{2}$ part wood ashes, $\frac{1}{8}$ part bone-meal and $\frac{1}{8}$ part ground chalk. Mix the ingredients together well, put into the hole, fill in evenly and tread down firmly. Give a good watering afterwards, and a week or so later plant the vine.

Buy two- or three-year-old canes and if they are to be grown on the cordon system, planting may be done as close as 3 ft. apart. If the vine is to be allowed to spread naturally, it may need 20 feet of room! Do the planting in November spreading the roots out carefully and shallowly and firming them in well.

General cultivation. Syringe the vine rod over in the morning and at night, and if the weather is hot, syringe round and about inside the house during the day as well so as to provide a humid atmosphere. Keep the temperature at night at 55°F. From the top of the rod a growth should develop strongly, and when it gets to a length of 8 ft. the tip of it should be cut off. This usually takes place about the beginning of September. In December this long growth should be cut back to within 6 ft. of soil level, just above a bud. From now on until the spring the rods are rested.

The rod as it grows should be tied up to the strong wires running parallel to the glass, but in January or early February the rods must be untied so as to allow the tops to bend down to ground level. They stay like this for a month or more so as to make certain that the buds along the whole length of the rod start to grow at the same time. Once all the buds are growing, tie the rod back to the wires once more, and in March give the border a good watering.

If two or three growths develop at each bud, rub out two of them in the early stages of the vine's life leaving only one to grow on. When these laterals are 1 ft. long cut off the tops, and if any side-growths develop as a result, cut these back at two leaves. Do not allow any fruit to form in the second year.

In the winter the side-growths that were allowed to develop should be cut back very hard – in fact back to one bud, and at the same time the end one-year-old growth or leader should be reduced by half. Once again the branch should be allowed to bend down and when all the buds have started to grow, should then be tied back.

In this third year the side-growths should fruit, but not more than eight bunches should be allowed per rod. In the summer prune as advised for the second year.

From the fourth year onwards, the pruning is standardised. The laterals are cut back in the winter to within one bud of their base, and the leader is reduced by half. To control the eggs of insect pests the rod should be painted with a tar-oil wash like Mortegg, making up a 10% solution. When the rod has reached the top of the house the leader will need cutting back almost entirely. As the rods get older it is advisable to remove the loose bark because this may be harbouring pests.

Each spring the laterals that develop should be stopped at two leaves beyond each bunch of fruit. The secondary laterals that develop can be cut back at just beyond the first leaf. Any tendrils that appear should be cut off at the same time.

Temperature and ventilation. Generally speaking, the temperature of a house should be about 50°F at night-time but may be allowed to rise to 70°F when the vine is in flower. When the little grapes have formed the temperature is lowered to 60°F so as to encourage them to stone properly, and a fortnight after this the temperature can again rise to 70°F. It is only when the grapes begin to colour that the temperature can be allowed to drop to 60°F again. In the winter, if there are no other plants in the house, there need be no heat at all.

There is perhaps no crop where automatic ventilators do more good than in the case of vines. Those who have to adopt hand ventilation should give a little air early in the morning, increase it gradually during the day, and then close down gradually in the afternoon until the ventilators are shut, say at 5.30 or 6 p.m. The exception is when the grapes start to ripen, in which case the ventilators must be open night and day.

Watering. Flood the vine border each winter – be prepared to give a lot of water just before flowering, and an equal quantity after thinning. In fact it pays to water well until the grapes start to ripen. Coupled with watering, spray the rods over early in the year to help the buds grow out freely; stop syringing when the vines come into flower and start again when the berries are forming, making certain to syringe the undersides of the leaves to keep down red spider. Stop syringing again when the grapes start to ripen.

Thinning. This is quite a big job and involves thinning out when the grapes are the size of sweet pea seeds. All the inner berries should be cut out, followed by the smaller berries and finally the side berries. A second thinning is usually necessary when the berries are the size of peas – it is surprising how many berries have to be cut off and it is best never to look downwards to see the quantities on the floor below! Those who do not thin rigorously never get nice bunches of big grapes and instead often get diseased shrivelled grapes, many of which rot off.

Varieties

Madresfield Court – a black muscat
Gros Colmar – large berries of excellent appearance, late, black
Alicante – a free fruiter, late, black
Muscat of Alexandria – a white, first-class quality, late
Buckland Sweetwater – early, handsome, white, fair flavour

PART FIVE

Greenhouse Pests and Diseases

INTRODUCTION

ONE of the most important factors in the control of pests and diseases is glasshouse hygiene. No amount of spraying, dusting and fumigating will produce clean, healthy crops if the houses are dirty and neglected.

Under the heading of hygiene must be included such measures as weed control, the immediate removal of rubbish from the house, the annual scrubbing down and spraying with disinfectant, fumigation, and the sterilisation of seed trays, pots, stakes and canes.

Weed control. Weeds must be kept down both inside and outside the houses, as they provide alternative host plants for pests and diseases attacking plants. For example, caterpillars of the tomato moth and other moths spend part of their life on nettles and docks; white fly and red spider can often be found on weeds, waiting to attack a greenhouse plant; aphides frequently breed on annual weeds. Regular cultivation should be carried out out of doors near the greenhouse, to avoid giving the weeds a chance to grow. All hoed-off weeds should be put immediately on the compost heap, and should not be left lying about. The gardener should avoid using poisonous weed-killers near the greenhouse as the chemical may be carried through into the house in the soil water. Rubbish of every kind, such as thinnings, side-shoots, leaves, stems and roots, should be removed and put on the compost heap.

Annual disinfection. Disinfection is usually carried out in the winter when the house may be emptied of plants. Spray the glass and woodwork thoroughly using a 1% solution of cresylic acid, and applying it with plenty of force so that the disinfectant penetrates into every crack and crevice. The brickwork or woodwork and staging should also be scrubbed down with the same solution. Particular attention should be paid to the underside of the staging and all rubbish cleared out from behind any heating pipes. A suitable spray can be made up using half a pint of cresylic acid, half a packet of a powder detergent such as Daz and 6 gallons of water.

Fumigation. Fumigation is carried out by many gardeners as a matter of routine, and when such measures are practised regularly serious pest and disease attacks are seldom experienced. For fumigation it is necessary to know the cubic air content of the house. This can be easily calculated for a span-roof house in the following way: measure the height to the eaves and the height to the ridge in feet, add these two figures together and divide by two; multiply the resulting figure by the length of the house in feet, and multiply this result again by the width of the house in feet. The result is the cubic capacity of the house. Thus, the capacity of a house measuring 4 ft. to the gutter and 12 ft. to the ridge, 12 ft. long and 10 ft. wide, is as follows:

the average height is 4 plus 12	=	16 ft.
divided by 2	=	8 ft.
multiplied by 12	=	96 ft.
multiplied by 10	=	960 cubic ft.

The house should be closed down completely before fumigation begins, and if there are any air spaces caused by badly fitting ventilators and doors, broken glass and so on, a little extra fumigant should be allowed. It is better, however, that general maintenance and repairs should be carried out in the house before fumigation, and badly fitting doors rectified and broken glass replaced.

The operation should, if possible, be carried out on a still evening, and with most fumigants better results are obtained when the temperature and humidity are high. It is necessary to keep the house tightly shut whilst fumigating until the following morning when the ventilators should be opened up wide. The most commonly used fumigants today are the 'smokes' obtainable from firms such as Messrs. Murphy of Wheathampstead, Herts. These consist of small canisters containing a specially prepared chemical lethal to a particular pest or group of pests, and it is only necessary to place them at intervals on the floor of the house, and set light to the fuse. This method saves a great deal of time and labour.

Spraying. In spraying it is essential to use the correct quantities, to mix thoroughly, to apply the wash at the right time so as to obtain maximum control of the pest, and to use a reliable spraying machine. Many pests and diseases are found on the undersides of the leaves, so particular attention should be paid to these. As a rule, the gardener should aim at covering every part of the plant with a fine mist-like spray, without actually drenching the plant; otherwise scorching may occur and, in addition, a lot of wash will be wasted.

Pests fall into two classes: those that suck sap from the inner tissues of the leaf and stem after piercing the outer membrane: the sucking pests; and those which actually eat the foliage: the biting pests. For the former a contact insecticide such as Derris or nicotine should be used, and the wash applied with plenty of force so that it actually comes into contact with the soft bodies of the insects. For the latter a stomach poison may be used, and this is best applied as a fine mist-like spray so that every part of the plant is covered and the insect takes some of the poisoned material into its mouth as it feeds.

The aerocide system. This recently devised system ensures a much more effective control of pests, as the insecticide is applied in minute particles and comes into intimate contact with insects, eggs, foliage and flowers. Application is by means of a special projector, consisting of a pressure container charged with specially prepared insecticide and liquid gas. When the pressure is released, the gas is liberated with considerable force, breaking up the liquid insecticide into minute droplets which are ejected into all parts of the glasshouse. Pests which can be controlled by this method include aphides, caterpillars, thrips, white fly and red spider.

Dusting. This is not usually regarded as being so efficient as spraying as a method of pest and disease control, since it is more difficult to distribute dusts evenly, and applications may have to be made several times in a season to obtain complete control. The essential is to obtain a good machine which will give efficient distribution. Dusting has, however, certain advantages: for example, in a dry season when water is short. Also it is quicker than spraying in many cases and its effect is less severe on seedlings and young plants.

CHAPTER 21

Insect Pests

ANTS. These sometimes make their nests in the roots of pot plants. In warm houses such as those used for growing orchids, they can be found in association with aphides, from which they obtain honeydew.

Control. The simplest method is to use Nippon ant killer.

APHIDES. Almost all glasshouse crops are subject to attack by black fly, greenfly, plant lice or aphides, as they are variously called. They frequently cause stunting and distortion of growth and malformed blooms. Also, the sticky honeydew which the insects secrete, and in which black moulds develop, renders plants unsightly and quite unfit for the market. Under glasshouse conditions these pests may continue to multiply throughout the winter, the wingless females often producing seven or eight young a day for three weeks or so.

Control. Fumigate regularly with nicotine shreds or cones. These are placed at intervals on the greenhouse floor, and a light is then set to them – do make sure that they merely smoke and do not flame.

CAPSID BUGS. These insects are especially a sucking pest of the chrysanthemums, but they also attack dahlias. Symptoms are growing shoots which become stunted and twisted, and sometimes 'blind'. The flowers are badly mis-shapen and small. The initial attack occurs when the plants are standing outside in the summer, but after they are housed in the autumn the depredations of the pest can become very severe.

The capsid attains a length of about ¼ in. when fully grown, and varies in colour from a pale green with a yellowish tinge to a pale grey-green. Adult insects hibernate outdoors on rubbish heaps and at hedge bottoms, but they can be found in glasshouses up to December. During the spring the overwintered adults begin laying eggs (which are creamy-white in colour) on the host plants. The female bugs actually lay their eggs right into the stem tissues of the plants. The young when hatched out are very active and pass through various stages of development before they reach maturity. The earliest hatched broods are normally fully developed by June and several generations can occur in one season.

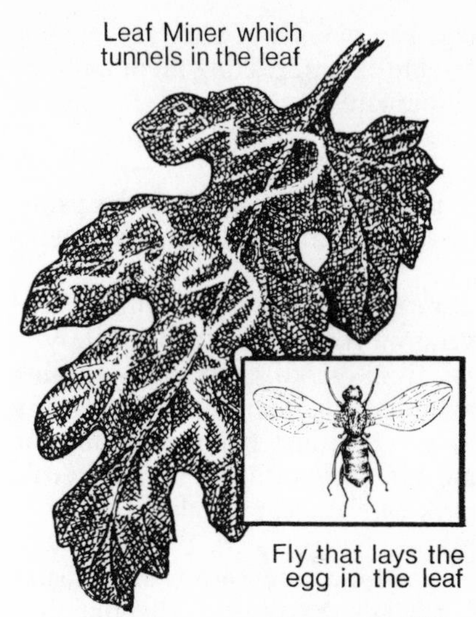

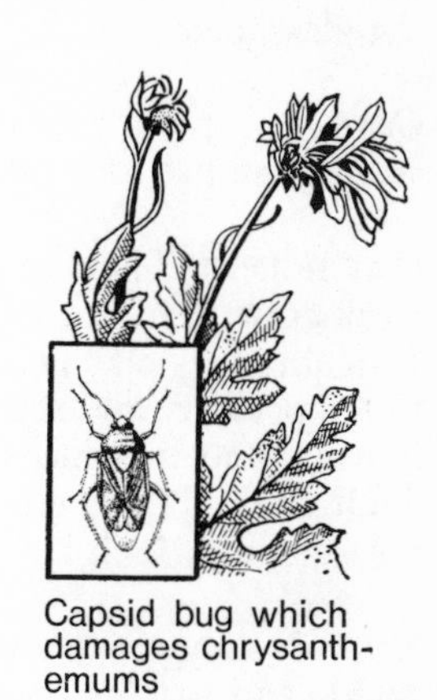

FIG. 29 Two insect pests

Control. All rubbish, old chrysanthemum stools and dahlia tubers, should be burnt and hedge bottoms cleaned out. When an attack has developed on a crop, nicotine will give a satisfactory kill. It can be applied as a spray, employing considerable pressure and a coarse nozzle, and directing the liquid well into the growing points of the plants. Nicotine fumigation is also effective, using shreds or proprietary tubes at the correct concentration.

CATERPILLARS. There are many types of caterpillars which attack glass-house plants. There is the tomato moth, which causes injury to flower crops such as chrysanthemums, as well as to tomatoes; the angle shades moth, which devours the leaves and flower buds of several glasshouse plants; and the carnation tortrix, which is also found on a variety of plants.

Eggs are laid in clusters, frequently on the undersides of the leaves. Hatching may take place in one to two weeks under glasshouse conditions, and the caterpillars feed voraciously for three or four weeks. The pupa or chrysalis stage may last anything from a few weeks to several months before the moth hatches.

Control. Apply Derris dust as soon as the caterpillars are seen, using a good duster and repeating the treatment every ten days or so until the pest is completely eradicated.

EARWIGS. Earwigs sometimes damage the blooms of chrysanthemums and dahlias under glass to a considerable extent, making them unfit for use. During the winter they may hibernate in cracks and crevices in the glasshouse.

Control. Apply a good Derris dust all round the house, on the paths, staging, and so on, and in any corners where the earwigs may hide.

LEAF HOPPER. These insects also suck sap from plants, causing stunted weak growth. Eggs are laid in the veins on the undersides of the leaves. Hatching may take place in anything from one to six weeks, according to temperature and season. The 'nymph' which emerges from the egg is white and translucent. It moults several times before reaching the adult stage. Plants attacked include the chrysanthemum, calceolaria, pelargonium, fuchsia, asparagus fern, salvia and primula.

Control. A thorough spraying with a good nicotine wash using a coarse nozzle and soaking the plants well will kill these pests. Nicotine dust applied early in the season is also satisfactory, but it should not be applied when the flowers are nearly ready to open, as it spoils their appearance. Fumigation with nicotine using $\frac{1}{4}$ oz. per 1000 cubic ft. can also be done.

LEAF MINER *Phytomyza atricornis.* The tiny grubs (of the Leaf Miner Fly) are about 2mm. long – greenish white in colour and legless. They tunnel into the leaves of plants in the greenhouse, and can be a serious pest of the chrysanthemums and cinerarias. The tunnelling is done in between the upper and lower surfaces of the leaf, leaving a snakey-like, whitish, yellow or brown trail which can easily be seen. They cross and recross as a rule. Badly attacked leaves will die.

Control. It is a simple matter to feel for the grub with finger and thumb which will be in one end of the tunnels, as a rule. Do this when the tunnelling is first seen. Kill the grub or grubs with the point of a knife and there will be no further trouble. Painting the leaves with Malathion may be done in a very serious case. Be very careful when handling this product.

LEATHERJACKETS. These are the maggots of the crane fly or daddy long-legs. They can be troublesome, especially in newly built houses on freshly used land. They damage the roots and stems of plants, and this can be particularly serious for young plants, causing them to wilt and die.

Eggs are laid early in the summer, and the maggots appear in about a fortnight. Whilst growing the maggots eat greedily and, as this stage lasts for six to eight months, considerable damage can be caused.

Control. Use a bait of Paris Green (1 lb.) or Bran (30 lb.) moistened with water and broadcast on the soil in the evening. *Very poisonous.*

MEALY BUG. Mealy bug is a common and troublesome pest on ferns, gardenias, asparagus fern and other greenhouse plants. The bug is covered with a white mealy wax and so is not easily controlled by sprays.

Clusters of eggs are laid in the leaf axils and at the bases of the leaves. They hatch in about a fortnight and the young bugs feed actively, sucking the sap from the leaves. In warm greenhouse conditions the life cycle is rapidly completed and breeding continues almost the whole year round.

Control. Routine measures must be carried out if mealy bugs are to be controlled successfully. This means first washing off the bugs with a soft brush dipped in methylated spirit. Sometimes old colonies can be washed off by holding the plant under a forceful jet of water from a tap or hose. After washing, a nicotine spray can be applied, using a coarse nozzle and thoroughly soaking the plants. Use half an ounce of nicotine in 5 gallons of water, with the addition of a good liquid detergent.

MILLIPEDES. These soil pests attack seedlings and gnaw the stems of plants just above ground level. They generally feed at night and burrow into the soil during the day.

These pests should not be confused with centipedes, which are beneficial. The latter may be distinguished by the fact that they have only one pair of legs on each segment, while the millipede has two.

Eggs are laid in the soil during the spring and early summer and hatch out in about a fortnight, the young millipedes feeding mostly on decaying organic matter. Two main species are found in the glasshouse: the common greenhouse millipede which has a flattened, segmented body, reddish-brown in colour and about one inch in length, and the spotted millipede which is only half an inch long, yellowish with a row of reddish or purple spots along each side. Some curl up like a watch-spring.

Control. Flaked naphthalene, applied at 4–8 oz. per square yard (the largest amount on heavy soil) and well watered in when the house is cleared of crops, will destroy them.

RED SPIDER. The red spider mite is one of the most serious of greenhouse pests, and amongst flower crops it attacks the carnation particularly, as well as arums, roses, salvias and asparagus fern. The tiny mites collect on the undersides of the leaves, and in a bad attack the foliage takes on a finely mottled appearance and may become greyish and shrivelled. A fine webbing appears on the leaves, and growth is poor and weak.

The mites usually hibernate over the winter in crevices in the greenhouse, also in canes, stakes, straw, and so on. Early in the new year the females start laying eggs, and from April till September breeding is almost continuous. The tiny eggs are laid on the undersides of the leaves, and hatching may take place in as little as three days in a warm house or up to ten days under cooler conditions. The young spider begins to feed immediately, and the whole life-cycle may take only eight or nine days.

Control. The latest method of controlling red spider under glass is by the use of azobenzene 'smokes'. These give complete control of the pest on all greenhouse crops, but its use is not advised on sweet peas, zinnias or schizanthus. The canisters merely have to be placed on the greenhouse floor and a light set to the fuse.

Fumigation with flaked naphthalene is an alternative. Two applications are usually necessary at intervals of about ten days. A high temperature (over 70°F) and a high humidity are necessary for successful results, and the quantity required is 1½ oz. per 100 cubic ft., according to the temperature and humidity, and the plants to be treated.

SCALE INSECTS. These are very similar in life history and habits to the mealy bug, but they are less active and feed in one place almost continuously. The young scales move about until they find a suitable feeding place, then they settle down and a hard leathery scale forms over their bodies. Eggs develop inside the female's body and, when the parent dies, remain protected by the scale until hatching takes place.

Control. As advised for mealy bug (see p. 247).

SLUGS. Slugs will attack almost any greenhouse crop devouring seedlings and the roots, shoots and foliage of older plants. They hide during the day in the soil, in damp, dark corners of the greenhouse, under boxes, pots, etc. and come out at night to feed.

Eggs are laid in clusters in damp places. They hatch in a month or so and the young slugs do little damage to crops at first, but later on start to feed voraciously. The commonest kind found under glass is the grey field slug – which may be yellowish, brown or purplish and is about 1½ in. long.

Control. One of the most important points is to keep the houses clean and free from rubbish and weeds in which the slugs can collect. Powdered copper sulphate mixed with an equal quantity of ground limestone may be forked into the soil at 4 oz. per 5 square yards when the house is empty of plants. If flower crops are badly attacked, Bordeaux mixture may be worked into the top soil, a light sprinkling being sufficient. This treatment may also be applied to plants in pots and boxes, scattering the mixture all round the pots and under the staging. 'Draza' pellets may also be sprinkled in the greenhouse. They are blue.

SYMPHYLIDS. These tiny white creatures are quite common in greenhouse soils, and are closely related to millipedes. Attacked plants frequently wilt due to damage to the roots, and may die or at least look sickly.

Eggs are laid singly or in small clusters in the soil, and hatch in two to three weeks. The young symphylids moult several times and reach maturity in five to six weeks. They are very difficult to see. They disappear quickly under soil crumbs on disturbance.

Control. Paradichlorbenzene may be applied to the soil in the spring at the rate of 1 oz. per yard run. This chemical will work more effectively if the soil is damp.

THRIPS. Thrips are sucking insects, and may do a great deal of damage in the greenhouse by sucking sap from leaves and flowers. Frequently flowers become so distorted and mottled that they are quite unfit for use. Plants attacked include the carnation, cyclamen, arum, azalea, cineraria, orchid, chrysanthemum and sweet pea.

There are several species of thrips, including the greenhouse thrips, the beet thrips, the onion thrips and the lily thrips. They are tiny black, brown or yellowish insects about one-tenth of an inch long. The white eggs are laid on the leaves and generally hatch in a week or two. The 'nymphs' which emerge attack the leaves, moulting about four times and finally emerging as the fully grown insect. Breeding and development are rapid and continue throughout the year under glasshouse conditions.

Control. The 'smokes' will also control thrips. A white oil or petroleum emulsion spray gives good control. Use a reliable proprietary brand and apply the spray as a fine mist with a good machine. It is important to wet both surfaces of the leaves, and to see that none of the spray accumulates in the crowns or leaf axils of the plants, otherwise damage may occur.

N.B. These oily sprays should not be used on flowering plants such as cyclamen, arums and carnations, or the flowers will be damaged. Also, the glaucous foliage of these plants would prevent such a spray from being fully effective. Naphthalene fumigation may be carried out where these crops are grown, grade 16 naphthalene being used at the rate of 1 oz. per 100 cubic ft., broadcast on the floor of the house in the late afternoon and the house kept closed until the following morning. It is important to carry out control measures against these insects as soon as they are seen as they are the carriers of virus diseases, the onion thrips in particular transmitting the dreaded tomato spotted wilt which attacks a variety of glasshouse plants.

WEEVILS. Grubs of the vine weevil are sometimes found feeding on the roots or crowns of pot plants such as cyclamen, pelargoniums, begonias, ferns and primulas.

Eggs are laid in the soil, hatching in two to three weeks, and the large white grubs feed throughout the summer and winter.

Control. Infested pot plants should be re-potted in fresh soil. Place a few crystals of benzene hexachloride in the bottom of the pot to kill the pests.

WHITE FLY. This insect attacks a wide range of glasshouse flower crops, including calceolaria, dahlia, freesia, cineraria, primula, coleus, arum, chrysanthemum and azalea. It sucks sap from the leaves and the loss of sap results in considerable weakening of the plant. A sticky substance called honeydew is exuded by the insects and this blocks up the pores of the leaves so that respiration is impeded. In addition a black sooty mould grows in the honeydew and spoils the appearance of the plant, sometimes to such an extent as to make it unusable.

Eggs are laid in groups on the undersides of the leaves, and these hatch in about a fortnight into 'scales' – flat oval bodies which immediately begin to feed on the leaf sap. These moult several times, and finally the mature fly emerges, a tiny white insect about $\frac{1}{25}$ in. long.

Control. A tiny parasitic fly, the chalcid 'wasp', lays its eggs in the white fly scales, and its grubs feed on the scales finally destroying them. The parasite can be obtained from the Royal Horticultural Society and when introduced into an infested greenhouse provides quite an effective means of control. Fumigation may be carried out periodically, roughly every two or three weeks, with a modern 'smoke'. This is a very simple method as it is only necessary to place the canisters along the floor of the house and set light to the fuse.

WIREWORMS. Wireworms, the grubs of the click beetle, are quite common under glass especially when virgin soil has been introduced into the house. Most greenhouse flower crops are subject to attack, the grubs feeding on the roots and sometimes causing the death of the plant.

Eggs are laid in the soil in July or August. The young grubs feed voraciously on roots, seeds and tubers in the soil, and this stage may last three or four years before pupation takes place. During the winter they burrow down several inches deep in the soil, and in the spring they return to the surface and start feeding again.

Control. A common method of wireworm control under glass is to trap them by inserting pieces of carrot or old cabbage stumps in the soil, the traps being placed between the rows of plants at intervals of about 2 ft. They should be inspected weekly and any wireworms removed and destroyed.

WOODLICE. These are often found in glasshouses, collected in cracks and crevices, under bricks, stones, boards, and so on. They feed on the stems, leaves and roots of crops.

Control. Derris or Pyrethrum dust, sprinkled on the soil all round the plants and in all the corners of the house, gives satisfactory control. This treatment may have to be repeated several times during the season.

Individual Plants and their Pests

Amaryllis

MEALY BUGS – see p. 247.
THRIPS – see p. 249.

Anthurium

APHIDES – see p. 244.
SCALE INSECTS – see p. 248.

Arum

THRIPS – see p. 249.
WHITE FLY – see p. 248.
RED SPIDERS – see p. 250.
APHIDES – the mottled arum aphis is one of the commonest of greenhouse aphides, attacking arums and cyclamen especially, but also many other plants. It is frequently found in or below the arum spathes, sucking sap from them and often making them unusable.

Control. See p. 244.

Asparagus Fern

RED SPIDERS – see p. 248.
SCALE INSECTS – see p. 248.
APHIDES – see p. 244.
LEAF HOPPERS – see p. 246.

Aspidistra

SCALE INSECTS – see p. 248.
THRIPS – see p. 249.

Azalea

APHIDES – see p. 244.
MEALY BUGS – see p. 247.
THRIPS – see p. 249.

LEAF MINER – the azalea leaf miner is a small moth with yellow and grey wings, having caterpillars which tunnel into the leaves, making ribbon-like channels. The leaves turn brown, taking on a scorched appearance, and may drop prematurely.

Eggs are laid on the undersides of the leaves near the veins. They hatch in about a week and the young caterpillars immediately begin to feed on the leaf tissue. When they are fully grown each caterpillar folds a leaf over a cocoon which it has built. Attacks are usually most prevalent during the winter months.

Control. Tetrachlorethane fumigation should be carried out as soon as the pest is noticed. A quarter pint of this chemical per 1000 cubic ft. should be used. A fairly dry atmosphere is necessary and a temperature of about 60°F suitable. The liquid should be sprinkled along the path of the house in the evening and no air admitted until the following morning. This treatment may have to be repeated several times to obtain complete control. The only drawback to the use of tetrachlorethane is the risk of scorching if the plants are in a mixed house.

TORTRIX MOTH – the vine tortrix sometimes attacks azaleas, the greyish or yellowish-green caterpillars spinning the leaves together with their webs and feeding on the foliage, thus spoiling the plants.

Control. See carnation tortrix, p. 253.

Begonia

EELWORM – the chrysanthemum eelworm sometimes attacks begonias (see chrysanthemum pests, p. 253) as also does the fern eelworm (see p. 256).

MITE – the Tarsonemid mite may attack begonias under glass causing the foliage to become very brittle and to curl up. The flowers may also be affected, the blooms frequently being distorted and the buds withered.

Control. As soon as the pest is seen the plants should be thoroughly dusted with fine sulphur dust. Frequent repetitions of this treatment are necessary. A nicotine or Derris spray such as is used against aphides also gives some measure of control.

Camellia

SCALE INSECTS – see p. 248.

Carnation

APHIDES – see p. 244.

THRIPS – see p. 249.

RED SPIDERS – see p. 248.

TORTRIX MOTH – attacks a variety of other greenhouse crops besides the carnation. The caterpillars roll the leaves up and spin them together to form a 'tent' in which they hide and feed. They feed usually on the growing points of the plants, the young shoots sometimes being completely destroyed. The flower-buds may also be attacked and the petals devoured.

Eggs are laid in clusters on the leaves, hatching out in two or three weeks into yellowish-green and very active caterpillars. They moult several times before spinning a web and pupating, the moth emerging in about a fortnight. Under glass, two generations are usually produced in a year, one in winter, and one in summer.

Control. This is a difficult pest to control, owing to its habit of sheltering in the leaves. A nicotine spray applied soon after the larvae hatch out and before they spin their webs gives satisfactory control. A forceful spray should be applied so that it penetrates into the leaf-clusters and shoot tips. Several applications during the season will be required for complete control.

Chrysanthemum

APHIDES – see p. 244.

EARWIGS – see p. 246.

CATERPILLARS – see p. 245.

LEAF HOPPERS – see p. 246.

CAPSIDS – see p. 244.

MIDGE – outbreaks of this serious pest occasionally occur, but it is not widespread. Galls are produced on the leaves and stems, growth is poor and stunted, and the flowers often deformed.

Eggs are laid on the young leaves, growing tips and round the flower buds. They hatch in one to two weeks into small white maggots which eat their way into the plant tissue, producing the typical galls. After three to four weeks they pupate inside the galls, and this stage lasts about ten days.

Control. This is difficult as the maggots spend most of their lives in the galls. A nicotine spray can be applied when eggs are being laid and when the midges emerge, and this must be repeated about once a week while the attack lasts. Pure nicotine should be used at the rate of 1 oz. per $12\frac{1}{2}$ gallons of water, with the addition of a good spreader, so that the plants are thoroughly wetted with the spray.

LEAF MINER – this also attacks cinerarias and other members of the Compositae family. White tunnels are made in the foliage which in a bad attack may become completely disfigured. The miner is a small, dark grey fly with large clear wings.

Eggs are laid singly on the undersides of the leaves, hatching in about a week into maggots which immediately tunnel into the leaves. After feeding for two or three weeks they pupate in the tunnels for one to three weeks. Under glass the insects continue to breed throughout the year.

Control. A nicotine spray, using 1 oz. of pure nicotine in $12\frac{1}{2}$ gallons of water with the addition of a spreader such as Estol H, gives good control. Both surfaces of the leaf should be well wetted, and the spraying must be carried out regularly, say at intervals of a fortnight. A nicotine dust will destroy the flies, as will fumigation with nicotine as carried out against aphides.

EELWORM – one of the worst pests of the chrysanthemum. The minute worm, which is invisible to the naked eye, affects the foliage, starting on the lower leaves and gradually working upwards. Dark patches first appear on the undersides of the leaves which wilt and turn brown and shrivelled. The whole plant is weakened and the flowers are frequently small and malformed. The production of cuttings is also affected. Large robust-growing varieties appear to be more susceptible than the smaller kinds. This pest has become increasingly widespread in recent years, and growers and suppliers of stock alike must pay more attention to its control if it is to be kept down.

Eggs are laid in the plant tissues and also in the axils of the leaves. The tiny eelworms which hatch out can move about in the films of moisture which cover the plant, and enter the internal tissues through the pores. Several generations are produced in a year as the whole life history only occupies about two weeks. The pest is easily transmitted in the soil moisture, on weeds, leaves of plants, tools, and so on.

Control. Once the pest is established in the plant no spray or fumigant can touch it. Warm water treatment of the stools is the only sure method of control, and this should be carried out as follows: after the stems have been cut down to within 4–6 in. of the base and any loose soil has been washed off, they should be immersed in warm water maintained at a temperature of 110°F for 20 minutes. At the end of this time the stools should be placed in clean cold water to cool off, then boxed up in sterilised soil.

ROOT MAGGOT – the maggots of the chrysanthemum fly sometimes attack the roots of greenhouse chrysanthemums, though attacks are not common. They make tunnels on the outsides of the roots and round the base of the stems, the roots crack open and growth is retarded.

Control. Naphthalene should be used as a repellent in the spring and summer, dusting it lightly round about the young plants at intervals of two or three weeks throughout the season. A solution of mercuric chloride watered on the beds, using 1 oz. of this very poisonous chemical in 10 gallons of water, will kill the maggots. This should be applied as soon as any signs of an attack are seen.

Cineraria

APHIDES – see p. 244.
LEAF MINERS – see chrysanthemum leaf miner, p. 254.
CATERPILLARS – see p. 245.
THRIPS – see p. 249.
WHITE FLY – see p. 250.

Coleus

FERN EELWORM – see p. 256.
WHITE FLY – see p. 250.

Croton

ANTS – see p. 244.
MEALY BUGS – see p. 247.
SCALE INSECTS – see p. 248.

Cyclamen

APHIDES – see p. 244.
MITES – see p. 248.
THRIPS – see p. 249.
WEEVILS – see p. 250.

Erica

SCALE INSECTS – see p. 248.

Ferns

APHIDES – see p. 244.
MEALY BUGS – see p. 247.
TORTRIX MOTH – see p. 253.
WEEVILS – see p. 250.

THRIPS – see p. 249.

EELWORM – also attacks begonias, gloxinias and orchids. The leaves become discoloured and the plants are generally poor, weak and unfit for use.

Control. Small ferns can be given warm water treatment as applied to chrysanthemum stools. Large plants should be regularly sprayed with a good nicotine wash, using ¼ oz. of pure nicotine in 3 gallons of water with the addition of a liquid detergent. This helps to check the pest though it does not completely control it. Care should be taken not to buy in any diseased plants, and dead leaves on infected plants should be removed and burned.

MITES – quite a common pest of glasshouse ferns. The mites are usually found on the young fronds from which they suck the sap, causing them to turn brown and become distorted and mis-shapen.

Eggs are laid on the uppersides of the fronds and on the scales of the rhizomes. They hatch very quickly, often in two or three days when the greenhouse temperature is high. The tiny white mites feed for a week or two, then have a resting period of a few days after which the adult mite emerges. Several generations are produced in a year.

Control. Fumigation with naphthalene as for red spider is usually effective. Sterilisation of young plants can be carried out by immersing them in warm water maintained at a temperature of 110°F for 20 minutes. It is important to use sterilised soil and clean pots, and to propagate only from healthy plants.

ROOT MEALY BUG – this also attacks other glasshouse plants but is probably most common on ferns. Sap is sucked from the roots, chiefly the outer ones, and the foliage turns a greyish colour, eventually wilting and dying off.

Eggs are laid in clusters round the roots, and the bugs hatch out in a week or two. In the warm glasshouse breeding continues practically all the year round.

Control. Infested roots should be dipped in a solution of nicotine, using ¼ oz. of nicotine in 3 gallons of water. The pots should be sterilised in hot water before re-potting and, of course, sterilised soil should be used. Alternatively, a 1% solution of nicotine may be watered on the soil all round the affected plants as soon as signs of an attack are seen.

Fuchsia

LEAF HOPPERS – see p. 246.

THRIPS – see p. 249.

Geranium

APHIDES – see p. 244.

CATERPILLARS – see p. 245.

LEAF HOPPERS – see p. 246.

Gloxinia

FERN EELWORM – see p. 256.

Hyacinth

APHIDES – see p. 244.

Hydrangea

APHIDES – see p. 244.

RED SPIDERS – see p. 248.

THRIPS – see p. 249.

Lily

THRIPS – the lily thrip is a serious pest, attacking chiefly lilies but also sometimes orchids. They live between the bulb scales and suck the sap, causing unhealthy growth, and sometimes the plants are killed. For the life history and control, see p. 249.

Orchids

ANTS – see p. 244.

APHIDES – see p. 244.

FERN EELWORM – see p. 256.

SCALE INSECTS – see p. 248.

THRIPS – see p. 249.

FUNGUS GNATS – the maggots of these small flies occasionally attack orchids. Eggs are often laid in the sphagnum moss of the orchid compost; the small white maggots feed on the roots and pseudo-bulbs and may do a considerable amount of damage. They pupate in cocoons in the moss or soil, and the mature flies quickly appear. The life history is rapidly completed and there may be several generations in one year.

Control. Routine fumigations with nicotine as for aphides will also control fungus gnats.

SPRINGTAILS – these little creatures sometimes attack orchids, feeding on the seedlings and young plants. They like damp conditions such as prevail in orchid houses, and may be found in decaying vegetation, moss, damp soil, and so on.

The creamy-white eggs are laid in damp places and quickly hatch out. The young springtails are very similar to the adults (the latter being dark brown or black), or occasionally white or greenish and covered with hairs.

Control. Very light fumigation as advised for aphides but using smaller quantities of nicotine, say ⅛ oz. per 1000 cubic ft., will control springtails. Dusting with Derris, sprinkling the powder lightly over the soil as soon as the pests are noticed, helps to repel them. This application may have to be repeated two or three times for complete control.

Primula

LEAF HOPPERS – see p. 246. WHITE FLY – see p. 250.
WEEVILS – see p. 250.

ROOT APHIDES – the auricula root aphis is often found on the roots of pot primulas under glass, and an attack may be quite serious. They suck sap from the roots causing the foliage to turn yellow and wilt.

These aphides reproduce viviparously, i.e. they do not lay eggs but produce living young, which are identical with the parents but smaller. Breeding may continue all the year round under greenhouse conditions.

Control. Treatment against this pest is similar to that for root mealy bug, see p. 256.

Rose

RED SPIDERS – see p. 248.

APHIS – commonly found on roses under glass especially on young plants. In the warm conditions of the glasshouse breeding is very rapid, and the plants quickly become covered with the large green aphides unless control measures are taken immediately the first signs of the pest are seen.

Control. As for other aphides, p. 244.

SCALE – this is a very common pest of roses under glass, and on older plants the branches may become covered with a scurvy mass of scales.

Control. As for other scale insects, see p. 248.

THRIPS – this pest is similar to other species of thrips, but it does not spend any of its life in the soil, remaining on the roses throughout the early stages of its life. Under glasshouse conditions they hibernate in the late autumn, spending the winter in cracks and crevices in the greenhouse.

Control. See p. 254.

Salvia

EELWORM – see p. 248.
LEAF HOPPERS – see p. 244.
RED SPIDERS – see p. 248.

Solanums, Schizanthus and Streptocarpus

APHIDES – see p. 244.

Tulips

APHIDES – see p. 244. The peach aphis and the potato aphis are important, as they are carriers of the virus which causes 'breaking' in tulips. See virus diseases, p. 246.

CHAPTER 22

Plant Diseases

Fungus Diseases

BOTRYTIS OR GREY MOULD. This is a very common fungus disease which will attack almost all glasshouse plants, particularly primulas, cyclamen, chrysanthemums and begonias. Symptoms are a typical grey fluffy mould, and leaf spotting and collar rot in some plants. In cyclamen the disease may also cause a spotting of the petals. It is most common under cool conditions, and where the atmosphere is damp and the plants crowded.

Control. Botrytis is difficult to control as the spores can live on dead organic matter in the soil such as straw and leaves, and the ordinary fungicides, such as copper and sulphur, do not affect it. There are proprietary botrytis dusts on the market which the gardener should apply early in the season as a preventive measure. A light dusting should be given to the seed bed. After planting out, the young plants can be dusted again, and this may be repeated a fortnight later. The best control, however, is that of prevention, by maintaining good cultural conditions and general hygiene.

MILDEW. Mildew attacks a variety of glasshouse plants, and is especially prevalent where cool, damp conditions prevail. The leaves and sometimes the stems become covered with a white powdery substance, and in bad attacks growth may be deformed and stunted.

Control. As much air as possible should be given to keep the atmosphere dry. Avoid splashing the foliage when watering. If an outbreak does occur, the plants should be dusted with a Karathane dust, using a good duster, as soon as the plants show any signs of an attack. Alternatively, a Thiovit spray can be applied.

RUST. Rust attacks many of the greenhouse plants including chrysanthemums, carnations and roses. Orange-brown patches appear on the undersides of the leaves and sometimes on the stems, and these turn black later in the season. The infected leaves frequently turn yellow and die or drop off.

Control. A Captan spray is the best control for rust. Spraying should commence as soon as the rusty patches are noticed and should be repeated every two or three weeks. A good spraying should be given to make sure that every part of the plant is covered with the fungicide.

DAMPING OFF. This is an all too common fungus disease which attacks seedlings in various stages. In some cases the seeds are attacked before emerging and therefore never germinate. Again the very young seedlings may succumb to the disease just as they come through the soil. The most typical form of attack, however, occurs when the seedlings topple over, collapse and die after they have germinated. The fungi spread rapidly and will soon cause the death of all the young plants in a seed box.

Control. It is essential that thin sowing should be practised as the fungi will more easily attack seedlings which are overcrowded, weak and spindly. Moreover, while the seed compost must be sufficiently moist to allow the seed to germinate quickly and to sustain growth afterwards, excessive moisture must be avoided as the fungi will flourish under damp stagnant soil conditions. It is also necessary to see that the drainage of seed pans and boxes is perfect. As a preventive Cheshunt compound should be watered on the seedlings, but it must be realised that this will not cure attacked seedlings, but only protect healthy ones from being attacked. Moreover, the damping-off fungi consists of several distinct varieties and some of them are not controlled by Cheshunt compound. In this case the only remedy is good cultural conditions and the quick disposal of any attacked plants. A sensible thing to do is to use Levington compost for seed boxes, and when watering to ensure that uncontaminated water is used as the disease can be carried in the water supply.

Individual Plants and their Diseases

Arum

ROOT ROT – the first symptoms are the yellowing of the leaves round the edges, followed by the whole leaf turning yellow and wilting. The tips of the flower-spathes turn brown and the spathes do not open properly. On examining the roots the gardener will find that the small side-roots are rotten at the tips, and the rot eventually spreads upwards to the corms.

Control. Any corms which show signs of rotting should be burned immediately, and the rest should be sterilised by immersing them for an hour in 2% formaldehyde solution, i.e. 1 pint of formaldehyde in 49 pints of water. Sterilised soil should be used as the disease organisms can live in the soil for long periods. The roots should also be given a thorough cleaning if an attack is experienced.

SOFT ROT – the tops of the corms are affected by this disease and the leaves frequently droop and die. The corm goes soft and mushy and emits an unpleasant smell. In a bad attack the rot may spread right to the roots, and the whole of the centre rot away leaving only the epidermis or outer skin. The bases of the leaves also turn dark and slimy, the flower-stalk topples over and the spathe turns brown.

Control. The necessary measures are exactly the same as those taken against root rot.

Begonia

BOTRYTIS – both leaves and flowers may be affected, becoming covered with the typical greyish mould. See p. 260.
MILDEW – see p. 260.

Carnation

RUST – some varieties of carnations are much more susceptible than others to rust. For control see p. 260.

LEAF SPOT – light brown spots with purple margins can be seen on the leaves, and tiny black specks in the centres which are the fruiting bodies of the fungus.

Control. Only healthy plants should be used for propagation as the disease is easily transmitted from plant to plant. A Captan spray, as recommended for rust, may be applied as soon as the first signs of the disease appear. With all sprays used on carnations a good spreader must be added as the waxy 'bloom' on the foliage makes it difficult to wet. Syringing should only be done on warm sunny days, and the gardener should avoid splashing the leaves when watering. Plenty of air should be given.

STEM ROT OR WILT – this is a very common disease amongst carnations, and it can be serious if a bad attack occurs. It may attack the young cuttings soon after they have been taken, or after potting up. The rot starts at the base of the stem and eventually completely girdles it, so that the plant wilts and dies.

Control. Propagation should be carried out only from healthy plants as the disease can easily be transmitted in the cuttings. Deep potting is a common cause of stem rot, so the gardener should try when potting to place the cuttings just deep enough to keep them upright in the compost, and avoid burying any more of the stem than is absolutely necessary. Water only when required and give plenty of ventilation.

Chrysanthemum

CROWN GALL – a bacterial disease which attacks the crowns of the plants causing swellings.

Control. Diseased plants should be pulled up and burned, and should not, of course, be used for propagation purposes. The soil should be sterilised (see p. 55).

LEAF SPOT – small, dark brown spots appear on the leaves, growing and merging into one another until finally the leaves curl up at the edges and drop off.

Control. All the diseased leaves should be removed and burnt. Karathane spray gives good control.

MILDEW – see p. 260.
RUST – see p. 260.

Cineraria

DAMPING OFF – see p. 261.

Cyclamen

STUNT – plants affected by this disease are generally unhealthy-looking, the leaves being small and the leaf stalks short. Also reddish-brown areas appear in the corms, and the flowers often open before the leaves appear.

Control. Sterilised soil should always be used and seed saved only from healthy plants.

Lily

BOTRYTIS – small orange spots appear on the leaves, gradually spreading and becoming covered with the typical grey mould. See p. 260 for control.

Narcissus

ROOT ROT – causes the base of the bulb and the roots to rot away, and in bad cases the whole bulb may rot. The flowers are malformed and the plant becomes stunted.

Control. The bulbs should be sterilised by soaking them for an hour in a solution of mercuric chloride using ¼ oz. of this chemical in 2 gallons of water. Mercuric chloride can be obtained in the form of tablets which are very easy to use – they merely have to be crushed and dissolved in water. This treatment should be carried out before planting, and the bulbs put into sterilised soil. Great care should be taken not to allow the bulbs to come into contact with any infection after they have been sterilised.

NEMATODE DISEASE – when the bulbs are cut across, a brown ring can be seen inside. The young leaves become twisted and speckled and turn yellow.

Control. Sterilise with hot water maintained at a temperature of 110°F for three hours. As soon as this time has elapsed the bulbs should be transferred to cold water, where they should be allowed to cool off thoroughly before planting in clean soil.

Rose

BLACK SPOT – a very common disease in which irregular black spots appear on the uppersides of the leaves. These grow and spread until, in a bad attack, almost the whole leaf may turn black. Premature leaf-fall occurs and the whole plant may be defoliated, with the result that growth is weakened and flowering the following year affected. Some varieties are much more susceptible than others.

Control. The fallen leaves should be collected and burned as the disease spores can live over the winter if the leaves are left lying about. A Captan spray should be used and plenty of air given to keep the atmosphere dry and buoyant.

STEM CANKER – this disease can be serious in glasshouse roses, whole branches often dying back and growth being generally poor and weak. Large cankered wounds develop on the stems starting as small reddish patches on the young shoots. Later, the patches spread and cracks develop, the canker sometimes completely girdling the branch and causing its death. Infection occurs through wounds made in pruning or in cutting the blooms.

Control. All infected parts should be cut out and burned immediately they are seen, and badly cankered bushes removed as infection is easily spread from diseased plants to healthy ones.

RUST – see p. 260.
CROWN GALL – see chrysanthemums, p. 263.

Sweet Peas

DOWNY MILDEW – the only leaf disease which attacks greenhouse sweet peas to any extent, young plants being frequently attacked. The foliage turns yellow and wilts, and the typical downy greyish-white mould can often be seen on the affected parts.

Control. Remove and burn all the infected plants, paying attention to ventilation and temperature in order to keep the atmosphere dryish and buoyant. Spray with Captan as soon as the trouble is seen.

STREAK – quite a common sweet pea disease thought to be caused by bacteria, but it may be a virus disease. Brown spots and streaks first appear near the base of the plant and gradually extend upwards until the leaf stalks, flowers and seed pods are all affected. The foliage wilts away and eventually the whole plant dies.

Control. The bacteria which cause streak are usually seed-borne. The most effective method of control is to sterilise the seed by immersing it in a 5% formaldehyde solution for five minutes.

Tulips

FIRE – caused by a species of botrytis, and very common on glasshouse tulips. The disease gets its name from the appearance of the attacked plants, which turn brownish and look as though they had been scorched. Yellow or brownish rot patches occur in the bulbs as well. The first signs of an attack are the tiny yellowish spots which appear on the leaves and stems. These grow slowly until the whole leaf is covered, and the typical grey mould of the botrytis fungus can be seen. Severely attacked flower-buds may never open at all, or if they do, white or brownish spots develop on the petals and make the flowers unfit for use.

Control. Infected plants should be removed and burned immediately. Careful handling of the bulbs is important as the disease attacks damaged bulbs more easily than sound ones. Any which show signs of rot should be discarded.

Virus Diseases

Unfortunately no complete control has yet been found for these serious diseases, and they are not yet fully understood by the scientists. A virus disease may be said to be an infection present in the plant juices, and the organisms responsible are so small as to be invisible even under a normal microscope.

Infection spreads very quickly from plant to plant, and there are various ways in which the diseases can be transmitted. The most usual are:

By the agency of sucking insects such as aphides and thrips, which carry the infected sap from one plant to another.

On the knives or fingers of a worker; the diseases can thus easily be spread during such operations as disbudding and picking.

Virus diseases may cause dwarfing, distortion of growth, streaking or blotching of the flowers and foliage, etc. One of the worst is spotted wilt, which almost describes itself, attacking almost every glasshouse flowering plant grown.

Control. No cure is yet known for an infected plant, so all that can be done is to pull it up and burn it. Sucking pests such as thrips should be controlled by routine spraying (see p. 249) and every effort should be made to prevent the disease being transmitted by human agency.

Virus Diseases Attacking Individual Plants

Chrysanthemum

YELLOWS – the first signs of an attack are the appearance of faint yellow stripes along the veins of the leaves, and later the whole leaf becomes pale yellowish-green. The young leaves are narrower than normal and stand erect.

ASPERMY – a virus which distorts the flowers.

Control. Remove and burn affected plants as soon as they are noticed.

Dahlias, Lilies, Roses and Sweet Peas

See Virus diseases, p. 265.

Tulips

BREAKING – causes the plants to become weak and the foliage to be mottled. The chief effect is on the flowers, which become variegated instead of the usual uniform colour. They are frequently streaked and speckled with other colours, mostly white.

Control. All diseased plants should be pulled up and burned immediately they are noticed. 'Broken' strains should never be used for propagation. Aphides should be kept down by regular spraying or fumigation.

PART SIX

Month by Month Reminders

IT is of course impossible to include in a single chapter absolutely every operation that may be necessary in the greenhouse. What has been done is to include all the important operations, and from time to time repeat the instructions – because obviously if you start to take chrysanthemum cuttings in January, you may easily be going on taking them until the end of March.

A certain amount of repetition is therefore clearly necessary. One great advantage of writing a book on the greenhouse is that instructions largely hold good both for the north and the south. Northerners may need to use a little more heat, and it is undoubtedly true that those who live, for example, round about Littlehampton in the south get more light and so can start things off a little earlier than a man who is doing his best in the Potteries.

The month has purposely been divided up into approximately two equal halves: 1st to 15th and 15th to the end of the month. It may be that in the spring the northerner will have to delay his operations a little because of lack of light; also, it is usual north of about Manchester to have a frost around 24th September, and plants like chrysanthemums will therefore have to be moved into the greenhouse earlier than in the south.

January

First half of month

Order in Levington or Alexpeat compost.
Order the pots required.
Wash and clean those used previously.
Clean the roof glass.
Pinch out growing points of schizanthus.
Take cuttings of carnations and chrysanthemums.
Cut back *Begonia* Lorraine when flowering is finished.
Put 'plunged' bulbs into a cold frame.
Sow sweet peas.
Sow seeds of *Asparagus plumosus* and *A. sprengeri*.

Pot on late sown stocks.
Feed primulas and water carefully.
Bring batches of bulbs into the greenhouse.

Second half of month

Pot on schizanthus.
Fumigate for greenfly if necessary.
Pot on calceolarias into 6 in. pots.
Split up ferns and pot up.
Sow tomato seed for earliest planting in heated greenhouse.

February

First half of month

Pick off faded blooms of cyclamen.
Take cuttings of Korean chrysanthemums.
Start achimenes tubers into growth.
Start begonia tubers on peat and sand.
Insert more chrysanthemum cuttings.
Pot on pelargoniums into 6 in. pots.
Sow seeds of gloxinias, salvias and coleus.
Bring dahlia tubers out of store and start them off.
Start clivias into growth.
Feed cinerarias that are growing well.
Sow seeds of *Primula obconica* and *Streptocarpus*.

Second half of month

Pot up rooted cuttings of geraniums.
Continue to propagate outdoor chrysanthemums.
Sow seeds of celosias and *Campanula pyramidalis*.
Start roots of caladiums in growth.
Pot up struck cuttings of late varieties of chrysanthemums.
Prune roses and fuchsias growing in pots.
Bring more bulbs in pots and bowls from frame into the greenhouse.
Take cuttings of crotons, dracaenas and aralias.
Sow sweet peas.
Sow seeds of cyclamen for flowering in 4 in. pots.

March

First half of month

Sow seeds of *Solanum capsicastrum* and *Primula kewensis*.

Make up hanging baskets if desired.
Start giving cacti and succulents a little water.
Start early-rooted carnations after they have made seven pairs of leaves.
Fumigate against greenfly.
Pot up cyclamen seedlings into 3 in. pots.

Second half of month

Pot on carnations into 6 in. pots.
Pot up orchids just commencing into new growth.
Start off dahlia tubers if you have not done this already.
Reduce the amount of water given to cyclamen.
Bulbs that have flowered may now be planted outside, leaves and all.
Sow seeds of *Impatiens* and *Celosia*.
Pot up achimenes, three tubers to a 6 in. pot.
Take basal cuttings of *Begonia* Lorraine.
Start crinums and hippeastrums into growth.
Be prepared to shade seedlings and tender plants.
Pick off flowers of *Azalea indica* and syringe overhead afterwards.
Pot-up rooted dahlia cuttings.
Bring pots of lilies into the greenhouse.
Stake and tie schizanthus.
Sow seeds of auriculas and large flowered petunias.
Sow tomatoes, cucumbers, melons.
Sow *Salvia splendens* seed if needed for bedding.
Stake tall-growing annuals.

April

First half of month

Prick-off all seedlings as necessary into Alexpeat compost.
Pot on all struck cuttings that are ready into 3 in. pots.
Take leaf cuttings of Rex begonias.
Remove dead fronds from ferns.
Syringe plants with water on bright days.
Sow seeds of annuals for greenhouse decoration and seeds of cinerarias for November flowering.
Bring pots of *Lilium regale* into the house.
Shade greenhouse if sun is bright.
Pot up chrysanthemums into 3 in. pots and stand in cold frame.
Pots of lachenalias must be kept moist until the foliage has died down.
Cut off dead flowers from genistas.
Sow seeds of saintpaulias.
Rest the arums gradually.

Second half of month

Pot up begonia tubers.
Stop the more forward chrysanthemums.
Sow more primulas.
Stand pot-grown roses, that have finished flowering, out of doors.
Sow seeds of smilax and grevillea.
Dry off gradually arum lilies as they finish flowering.
Start giving ample ventilation to prevent scorching.
Dry cyclamen off gradually as they cease blooming.
Re-pot azaleas into Alexpeat compost.
Take cuttings from euphorbias.
Sow seeds of cucumbers, French beans and runners for outdoor planting.

May

First half of month

Snip off faded flowers of genistas.
Syringe ferns.
Pot off fuchsias, geraniums and heliotropes.
Pinch out growing points of fuchsias to encourage branching.
Pot on young cyclamens into 5 in. pots.
Take cuttings of poinsettias.
Place arum lilies outside in a sheltered position.

Second half of month

Keep nerines moist at roots until the foliage dies down.
Prick off cineraria and primula seedlings.
Take cuttings of eupatoriums.
Do final potting of the perpetual flowering carnations.
May be warm enough now to leave ventilators open throughout the night.
May also be possible to cut off the heat.
Re-pot acacias, heathers, and pot on celosias and gloxinias.

June

First half of month

Remove from greenhouse some winter occupants like azaleas, acacias, camellias, and *Solanum capsicastrum*; the last should be put in a frame.
Put cyclamen corms which have finished flowering in cold frame.

Pot up scarlet salvias into 3 in. pots.
Do the final potting of the late chrysanthemums into 8 in. pots.
Watch out for red spider on perpetual flowering chrysanthemums.
Prune climbing roses the moment they pass out of bloom.
Water tomatoes freely now and keep the plants well tied up.
Fertilise female flowers of melon.
Get all bedding material out of greenhouse to provide more room.
Feed begonias, gloxinias, hydrangeas, carnations and fuchsias with Marinure.
Pot up a batch of early freesias.
Take cuttings of the best coleus plants.
Sow seeds of cinerarias and humeas.
Prune back the brooms after flowering.

Second half of month

Seedling primulas should be hardened off in a cold frame.
Attend to the staking of chrysanthemums in pots and do any stopping necessary.
When hydrangeas pass out of bloom prune back the growths quite hard.
Sow cineraria again.
Feed fast-growing tomatoes after two trusses have set.

July

First half of month

Pot up the baby gladioli.
Sow seeds of gloxinias and stocks.
Young cyclamen plants may be ready for potting into 6 in. pots.
Regal pelargoniums which have finished flowering can be cut back hard.
Knock pot roses out of pots and re-pot in Alexpeat compost No. 2.
Pot up poinsettia cuttings struck a few weeks ago.
Sow seeds of eucalyptus.
With vines the atmosphere must be allowed to become a little drier, but not the soil.
When gloxinias pass out of flower, stand the pots in the cold frame.
Tuberous-rooted begonias may need staking.
Sow seeds of mignonette, double wallflowers, and late calceolarias to flower in winter.

Second half of month

Watch out for pests and diseases of chrysanthemums, and spray accordingly.
Remove all flowers from winter flowering begonias.
Sow seeds of double-flowered nasturtiums.
Continue to stake and disbud carnations.
Start to pot up early bulbs like Roman hyacinths.
Pinch out the tops of early tomatoes and feed with Marinure.
Cut back lateral growths of cucumber.

August

First half of month

Feed chrysanthemums regularly.
Pot up first batch of freesias and lachenalias.
Cyclamen corms which have been resting may now be re-potted.
Arum lilies that are resting outside may be given some water, and in a week or so may be re-potted.
As achimenes pass out of flower, transfer pots to cold frame.
Pot up bulbs of narcissus for flowering in winter and spring.
Pot up hyacinths and early flowering tulips.
Make a sowing of winter and spring flowering stocks.
Top-dress and feed cucumber beds.
Keep melons on the dry side if fruits are ripening.
Feed chrysanthemums in pots.

Second half of month

Sow seeds of nemesias and godetia, to grow on as pot plants.
There is still time to sow cyclamen for flowering the winter after next.
Take a few fuchsia cuttings and keep them growing steadily during the winter.
Keep poinsettias sturdy by keeping temperature at 60°F and yet giving plenty of ventilation.
Sow mustard and cress weekly if needed.
Feed chrysanthemums periodically with Marinure.
Gradually withhold water from hippeastrums.
If shading has been sprayed on the greenhouses, gradually wash it off.
Sow seeds of clarkias and salpiglossis, for pot plants.
Be sure to replace any broken panes before September.
Re-pot the francoas and scarlet nerines.
Pot-up the bulbs of the vallota.

September

First half of month

Trim leaves of tomato plants with the aim of ripening the fruits and clearing the crop this month ready for chrysanthemums.
Keep poinsettias reasonably near roof glass to keep the plants short-jointed.
Pot up another batch of bulbs if necessary.
Make a sowing of cyclamen seeds, which should bloom in 15 months.
Take cuttings of abutilons.
Pot up a batch of freesias for a late display.
Look around the greenhouse and discard all plants of no further use.
Bring in pots of arum lilies, re-pot or top-dress.
If necessary, do the first potting of primulas and cinerarias.

Second half of month

Watch out for white fly and fumigate.
Bring into the greenhouse azaleas, genistas and regal pelargoniums.
Give gloxinias water only when actually required.
Remove all shading from the house.
Ventilate as much as possible, and water regularly except in the case of plants that are going to rest like hippeastrums, gloxinias and begonias.
Watch out for the leaf miner maggot and spray with Malathion.
Pot up more dwarf early flowering gladioli corms.
In the north get ready to bring chrysanthemums into the greenhouse before the first frost.
Indian azaleas may be brought inside before the end of the month in the south.
Continue to disbud the chrysanthemums and watch out for pests and diseases, using the appropriate sprays, when necessary.

October

First half of month

Pot up African tuberose bulbs.
Stake and tie winter flowering begonias and browallias.
Continue to disbud chrysanthemums.
Keep growing freesias close to the glass on the shelf.
Pot up the most forward cinerarias into 6 in. pots.
Pot up a late batch of hyacinths and tulips.
Propagate perpetual flowering carnations by cuttings.

Pot up Canterbury bells into 6 in. pots.

Get rid of surplus coleus, keeping one or two of the best as stock plants.

A little fire heat will undoubtedly be necessary at night-time to keep out the frost. If it is very severe the tops of the chrysanthemums may be protected with sheets of paper at night.

Sow a second batch of cyclamen seeds.

Second half of month

Fuchsias may be put under the staging of the greenhouse, as may hydrangeas and crinums.

Pot up *Lilium regale* carefully and plunge pot outside in ashes with some protection to prevent excessive rain.

Pot on young specimens of humea.

Do not allow atmosphere in the greenhouse to become cold and damp.

Pot up retarded crowns of lily-of-the-valley.

June-sown primulas and calceolarias can be potted up into 6 in. pots.

Throw away the old plants of streptocarpus and let the young specimens take their place.

Bring the primulas from the cold frame into the greenhouse.

Be very careful when watering *Primula malacoides* not to let water enter clusters of leaves.

Bring some solanums into the warm to start colouring the berries.

Prune the climbers.

Keep down mildew.

Pot up *Lilium longiflorum* bulbs for blooming at Easter.

November

First half of month

Keep the air buoyant for the chrysanthemums.

Reduce the watering to the cacti and succulents.

Pot up bulbs of amaryllis.

Feed cyclamen and primulas in bud with Marinure.

Bring Paper White narcissii into the greenhouse where they will soon flower.

Continue to disbud late varieties of chrysanthemums.

Pot up *Lilium harrisi.*

Second half of month

When some varieties of chrysanthemums pass out of flower cut them down hard and place pots in frame for cutting production.

Fumigate the house against aphides.

Stake freesias and lachenalias.

Keep hydrangeas slightly on the dry side for the next three or four months.

Bring the clivias into flower – give water liberally, and more warmth.

Pot up spiraeas, kalmias and viburnums for forcing.

Bring into the greenhouse the potted Solomon's seal and potted lily-of-the-valley roots.

Avoid excessive heat this month – open the ventilators whenever possible.

December

First half of month

Bring in more bulbs in bowls and pots from outside.

Prune back the climbing plumbagoes fairly hard.

Look out for pests and spray.

As the chrysanthemums are cut down, select certain plants for cutting production.

Commence the propagation of perpetual flowering carnations.

Remove dead leaves and decayed blooms regularly.

Thin out growths of climbing lapagerias.

Second half of month

Spray with nicotine if necessary.

Prune the pot-grown roses and bring into the greenhouse.

Put solanums into the warmest corner of the greenhouse where there is plenty of light to ripen up berries.

Wash the glass of the greenhouse outside and lime-wash the walls inside.

Wash pots; order composts; paint labels.

Prepare and mend seed boxes.

Put the first batch of rooted carnation cuttings on a shelf near the glass.

Pot up a clump or two of dicentras for forcing.

Be careful not to over-water annuals this month.

Make a list of all bulbs, tubers, seeds and corms that will be required next year.

Temperature Conversion

Celsius (Centigrade) temperatures in the text have been confined to increments of half a degree and this table provides a cross reference to the Fahrenheit scale on that basis.

°C	°F	°C	°F	°C	°F	°C	°F	°C	°F
-1.0	30	19.0	66	39.0	102	59.5	139	80.0	176
-0.5	31	19.5	67	39.5	103	60.0	140	80.5	177
0.0	32	20.0	68	40.0	104	60.5	141	81.0	178
0.5	33	20.5	69	40.5	105	61.0	142	81.5	179
1.0	34	21.0	70	41.0	106	61.5	143	82.0	180
1.5	35	21.5	71	41.5	107	62.0	144	82.5	180.5
2.0	36	22.0	72	42.0	108	62.5	144.5	83.0	181
2.5	36.5	22.5	72.5	42.5	108.5	63.0	145	83.5	182
3.0	37	23.0	73	43.0	109	63.5	146	84.0	183
3.5	38	23.5	74	43.5	110	64.0	147	84.5	184
4.0	39	24.0	75	44.0	111	64.5	148	85.0	185
4.5	40	24.5	76	44.5	112	65.0	149	85.5	186
5.0	41	25.0	77	45.0	113	65.5	150	86.0	187
5.5	42	25.5	78	45.5	114	66.0	151	86.5	188
6.0	43	26.0	79	46.0	115	66.5	152	87.0	189
6.5	44	26.5	80	46.5	116	67.0	153	87.5	189.5
7.0	45	27.0	81	47.0	117	67.5	153.5	88.0	190
7.5	45.5	27.5	81.5	47.5	117.5	68.0	154	88.5	191
8.0	46	28.0	82	48.0	118	68.5	155	89.0	192
8.5	47	28.5	83	48.5	119	69.0	156	89.5	193
9.0	48	29.0	84	49.0	120	69.5	157	90.0	194
9.5	49	29.5	85	49.5	121	70.0	158	90.5	195
10.0	50	30.0	86	50.0	122	70.5	159	91.0	196
10.5	51	30.5	87	50.5	123	71.0	160	91.5	197
11.0	52	31.0	88	51.0	124	71.5	161	92.0	198
11.5	53	31.5	89	51.5	125	72.0	162	92.5	198.5
12.0	54	32.0	90	52.0	126	72.5	162.5	93.0	199
12.5	54.5	32.5	90.5	52.5	126.5	73.0	163	93.5	200
13.0	55	33.0	91	53.0	127	73.5	164	94.0	201
13.5	56	33.5	92	53.5	128	74.0	165	94.5	202
14.0	57	34.0	93	54.0	129	74.5	166	95.0	203
14.5	58	34.5	94	54.5	130	75.0	167	95.5	204
15.0	59	35.0	95	55.0	131	75.5	168	96.0	205
15.5	60	35.5	96	55.5	132	76.0	169	96.5	206
16.0	61	36.0	97	56.0	133	76.5	170	97.0	207
16.5	62	36.5	98	56.5	134	77.0	171	97.5	207.5
17.0	63	37.0	99	57.0	135	77.5	171.5	98.0	208
17.5	63.5	37.5	99.5	57.5	135.5	78.0	172	98.5	209
18.0	64	38.0	100	58.0	136	78.5	173	99.0	210
18.5	65	38.5	101	58.5	137	79.0	174	99.5	211
				59.0	138	79.5	175	100.0	212

Index

WITHDRAWN
HIGHLAND
REGIONAL
LIBRARY